www.rbooks.co.uk

ALSO BY STEPHEN HAWKING

A Brief History of Time
A Briefer History of Time
Black Holes and Baby Universes and Other Essays
The Illustrated A Brief History of Time
The Universe in a Nutshell

FOR CHILDREN

George's Secret Key to the Universe (with Lucy Hawking)
George's Cosmic Treasure Hunt (with Lucy Hawking)

ALSO BY LEONARD MLODINOW

A Briefer History of Time
The Drunkard's Walk: How Randomness Rules Our Lives
Euclid's Window: The Story of Geometry from Parallel Lines to Hyperspace
Feynman's Rainbow: A Search for Beauty in Physics and in Life

FOR CHILDREN

The Last Dinosaur (with Matt Costello)
Titanic Cat (with Matt Costello)

THE GRAND DESIGN

STEPHEN HAWKING
AND LEONARD MLODINOW

BANTAM PRESS
LONDON · TORONTO · SYDNEY · AUCKLAND · JOHANNESBURG

TRANSWORLD PUBLISHERS
61–63 Uxbridge Road, London W5 5SA
A Random House Group Company
www.rbooks.co.uk

First published in Great Britain in 2010 by Bantam Press
an imprint of Transworld Publishers

Book design by Simon M. Sullivan

A CIP catalogue record for this book is available from the British Library.

ISBNs 9780593058299 (cased)
9780593058305 (tpb)

Addresses for Random House Group Ltd companies outside the UK can be found at: www.randomhouse.co.uk
The Random House Group Ltd Reg. No. 954009

Printed and bound in India by Replika Press Pvt. Ltd.
4 6 8 10 9 7 5 3

CONTENTS

THE GRAND DESIGN

1

THE MYSTERY OF BEING

We each exist for but a short time, and in that time explore but a small part of the whole universe. But humans are a curious species. We wonder, we seek answers. Living in this vast world that is by turns kind and cruel, and gazing at the immense heavens above, people have always asked a multitude of questions: How can we understand the world in which we find ourselves? How does the universe behave? What is the nature of reality? Where did all this come from? Did the universe need a creator? Most of us do not spend most of our time worrying about these questions, but almost all of us worry about them some of the time.

Traditionally these are questions for philosophy, but philosophy is dead. Philosophy has not kept up with modern developments in science, particularly physics. Scientists have become the bearers of the torch of discovery in our quest for knowledge. The purpose of this book is to give the answers that are suggested by recent discoveries and theoretical advances. They lead us to a new picture of the universe and our place in it that is very different from the traditional one, and different even from the picture we might have painted just a decade or two ago. Still, the first sketches of the new concept can be traced back almost a century.

According to the traditional conception of the universe, objects move on well-defined paths and have definite histories. We can specify their precise position at each moment in time. Although that account is successful enough for everyday purposes, it was found in the 1920s that this "classical" picture could not account for the seemingly bizarre behaviour observed on the atomic and

". . . And *that* is my philosophy."

subatomic scales of existence. Instead it was necessary to adopt a different framework, called quantum physics. Quantum theories have turned out to be remarkably accurate at predicting events on those scales, while also reproducing the predictions of the old classical theories when applied to the macroscopic world of daily life. But quantum and classical physics are based on very different conceptions of physical reality.

Quantum theories can be formulated in many different ways, but what is probably the most intuitive description was given by Richard (Dick) Feynman, a colourful character who worked at the California Institute of Technology and played the bongo drums at a strip joint down the road. According to Feynman, a system has not just one history but every possible history. As we seek our answers, we will explain Feynman's approach in detail, and employ it to explore the idea that the universe itself has no single history, nor even an independent existence. That seems like a radical idea, even to many physicists. Indeed, like many notions in today's science, it appears to violate common sense. But common sense is based upon everyday experience, not upon the universe as it is revealed through the marvels of technologies such as those that allow us to gaze deep into the atom or back to the early universe.

Until the advent of modern physics it was generally thought that all knowledge of the world could be obtained through direct observation, that things are what they seem, as perceived through our senses. But the spectacular success of modern physics, which is based upon concepts such as Feynman's that clash with everyday experience, has shown that that is not the case. The naive view of reality therefore is not compatible with modern physics. To deal with such paradoxes we shall adopt an approach that we call model-dependent realism. It is based on the idea that our brains interpret the input from our sensory organs by making a model of the world. When such a model is successful at explaining events, we tend to attribute to it, and to the elements and concepts that constitute it, the quality of reality or absolute truth. But there may be different ways in which one could model the same physical situation, with each employing different fundamental elements and concepts. If two such physical theories or models accurately predict the same events, one cannot be said to be more real than the other; rather, we are free to use whichever model is most convenient.

In the history of science we have discovered a sequence of better and better theories or models, from Plato to the classical theory of Newton to modern quantum theories. It is natural to ask: Will this sequence eventually reach an end point, an ultimate theory of the universe, that will include all forces and predict every observation we can make, or will we continue forever finding better theories, but never one that cannot be improved upon? We do not yet have a definitive answer to this question, but we now have a candidate for the ultimate theory of everything, if indeed one exists, called M-theory. M-theory is the only model that has all the properties we think the final theory ought to have, and it is the theory upon which much of our later discussion is based.

M-theory is not a theory in the usual sense. It is a whole family of different theories, each of which is a good description of observations only in some range of physical situations. It is a bit like a map. As is well known, one cannot show the whole of the earth's surface on a single map. The usual Mercator projection used for maps of the world makes areas appear larger and larger in the far north and south and doesn't cover the North and South Poles. To faithfully map the entire earth, one has to use a collection of maps, each of which covers a limited region. The maps overlap each other, and where they do, they show the same landscape. M-theory is similar. The different theories in the M-theory family may look very different, but they can all be regarded as aspects of the same underlying theory. They are versions of the theory that are applicable only in limited ranges—for example, when certain quantities such as energy are small. Like the overlapping maps in a Mercator projection, where the ranges of different versions overlap, they predict the same phenomena. But just as there is no flat map that is a good representation of the earth's entire surface, there is no single theory that is a good representation of observations in all situations.

We will describe how M-theory may offer answers to the question of creation. According to M-theory, ours is not the only universe. Instead, M-theory predicts that a great many universes were created out of nothing. Their creation does not require the intervention of some supernatural being or god. Rather, these multiple

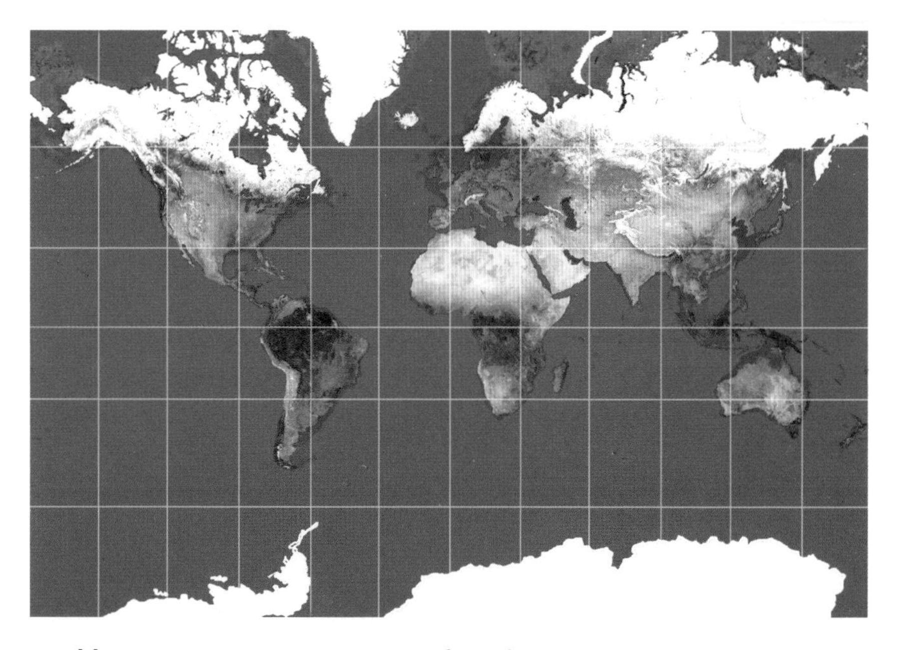

World Map It may require a series of overlapping theories to represent the universe, just as it requires overlapping maps to represent the earth.

universes arise naturally from physical law. They are a prediction of science. Each universe has many possible histories and many possible states at later times, that is, at times like the present, long after their creation. Most of these states will be quite unlike the universe we observe and quite unsuitable for the existence of any form of life. Only a very few would allow creatures like us to exist. Thus our presence selects out from this vast array only those universes that are compatible with our existence. Although we are puny and insignificant on the scale of the cosmos, this makes us in a sense the lords of creation.

To understand the universe at the deepest level, we need to know not only *how* the universe behaves, but *why*.

Why is there something rather than nothing?
Why do we exist?
Why this particular set of laws and not some other?

This is the Ultimate Question of Life, the Universe and Everything. We shall attempt to answer it in this book. Unlike the answer given in *The Hitchhiker's Guide to the Galaxy*, ours won't be simply "42".

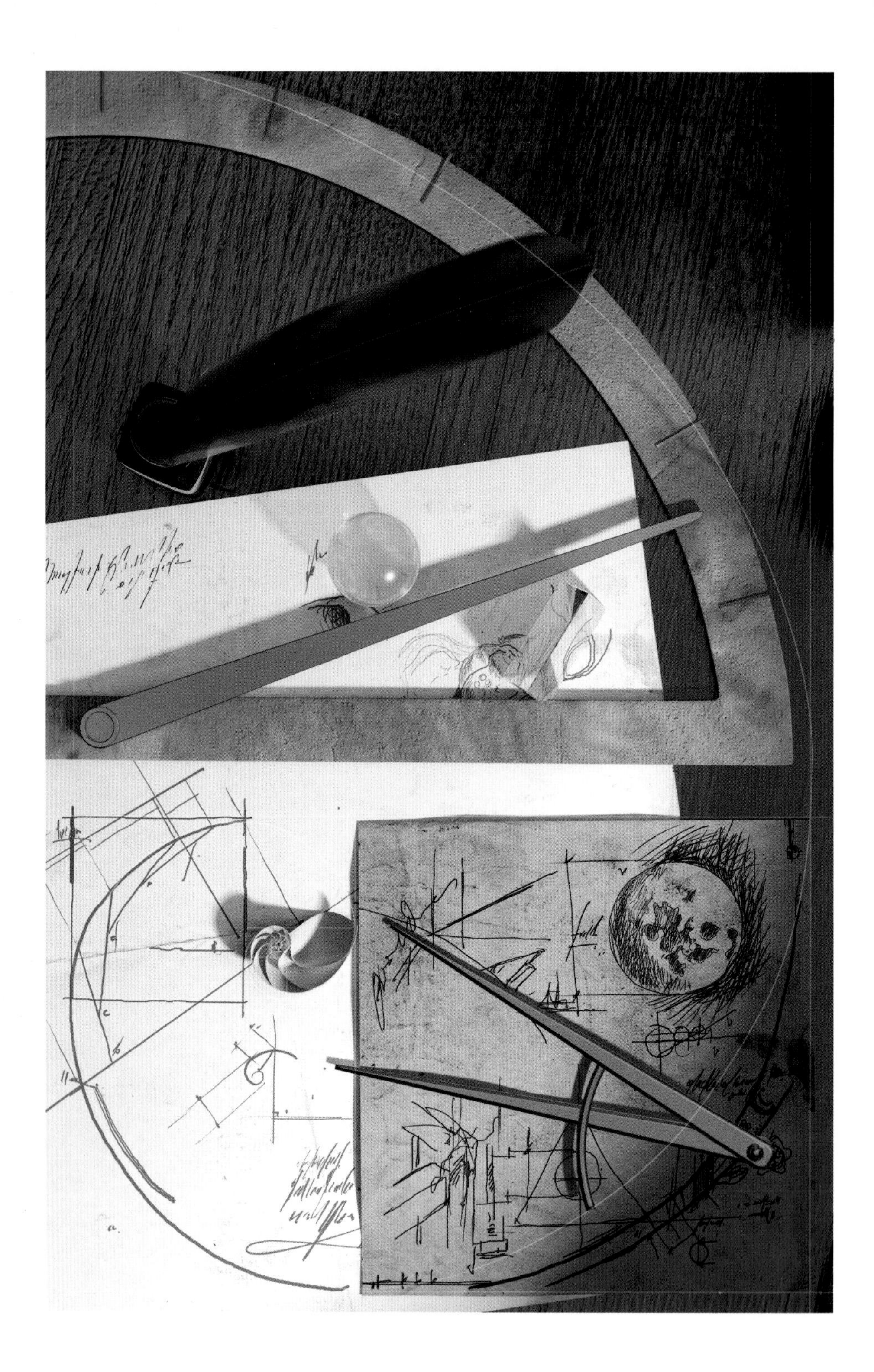

2

THE RULE OF LAW

Skoll the wolf who shall scare the Moon
Till he flies to the Wood-of-Woe:
Hati the wolf, Hridvitnir's kin,
Who shall pursue the sun.

—"GRIMNISMAL", *The Elder Edda*

IN VIKING MYTHOLOGY, Skoll and Hati chase the sun and the moon. When the wolves catch either one, there is an eclipse. When this happens, the people on earth rush to rescue the sun or moon by making as much noise as they can in hopes of scaring off the wolves. There are similar myths in other cultures. But after a time people must have noticed that the sun and moon soon emerged from the eclipse regardless of whether they ran around screaming and banging on things. After a time they must also have noticed that the eclipses didn't just happen at random: they occurred in regular patterns that repeated themselves. These patterns were most obvious for eclipses of the moon and enabled the ancient Babylonians to predict lunar eclipses fairly accurately even though they didn't realize that they were caused by the earth blocking the light of the sun. Eclipses of the sun were more difficult to predict because they are visible only in a corridor on the earth about 30 miles wide. Still, once grasped, the patterns made it clear the eclipses were not dependent on the arbitrary whims of supernatural beings, but rather governed by laws.

Despite some early success predicting the motion of celestial

Eclipse The ancients didn't know what caused eclipses, but they did notice patterns in their occurrence.

bodies, most events in nature appeared to our ancestors to be impossible to predict. Volcanoes, earthquakes, storms, pestilences and ingrown toenails all seemed to occur without obvious cause or pattern. In ancient times it was natural to ascribe the violent acts of nature to a pantheon of mischievous or malevolent deities. Calamities were often taken as a sign that we had somehow offended the gods. For example, in about 5600 BC the Mount Mazama volcano in Oregon erupted, raining rock and burning ash for years, and leading to the many years of rainfall that eventually filled the volcanic crater today called Crater Lake. The Klamath Indians of Oregon have a legend that faithfully matches every geologic detail of the event but adds a bit of drama by portraying a human as the cause of the catastrophe. The human capacity for guilt is such that people can always find ways to blame themselves. As the legend goes, Llao, the chief of the Below World, falls in love with the beautiful human daughter of a Klamath chief. She spurns him, and in revenge Llao tries to destroy the Klamath with fire. Luckily, according to the legend, Skell, the chief of the Above World, pities the humans and does battle with his underworld counterpart. Eventually Llao, injured, falls back inside Mount Mazama, leaving a huge hole, the crater that eventually filled with water.

Ignorance of nature's ways led people in ancient times to invent gods to lord it over every aspect of human life. There were gods of love and war; of the sun, earth and sky; of the oceans and rivers; of rain and thunderstorms; even of earthquakes and volcanoes. When the gods were pleased, mankind was treated to good weather, peace and freedom from natural disaster and disease. When they were displeased, there came drought, war, pestilence and epidemics. Since the connection of cause and effect in nature was invisible to their eyes, these gods appeared inscrutable, and people at their mercy. But with Thales of Miletus (c. 624 BC–c. 546 BC) about 2,600 years ago, that began to change. The idea arose that nature follows consistent principles that could be deciphered. And so began the long process of replacing the notion of the reign of gods with the concept of a universe that is governed by laws of nature, and created according to a blueprint we could someday learn to read.

Viewed on the timeline of human history, scientific inquiry is a very new endeavour. Our species, *Homo sapiens*, originated in sub-Saharan Africa around 200,000 BC. Written language dates back only to about 7000 BC, the product of societies centred around the cultivation of grain. (Some of the oldest written inscriptions concern the daily ration of beer allowed to each citizen.) The earliest written records from the great civilization of ancient Greece date back to the ninth century BC, but the height of that civilization, the "classical period", came several hundred years later, beginning a little before 500 BC. According to Aristotle (384 BC–322 BC), it was around that time that Thales first developed the idea that the world can be understood, that the complex happenings around us could be reduced to simpler principles and explained without resorting to mythical or theological explanations.

Thales is credited with the first prediction of a solar eclipse in 585 BC, though the great precision of his prediction was probably a lucky guess. He was a shadowy figure who left behind no writings of his own. His home was one of the intellectual centres in a region called Ionia, which was colonized by the Greeks and exerted an influence that eventually reached from Turkey as far west as Italy. Ionian science was an endeavour marked by a strong interest in uncovering fundamental laws to explain natural phenomena, a tremendous milestone in the history of human ideas. Their approach was rational and in many cases led to conclusions surprisingly similar to what our more sophisticated methods have led us to believe today. It represented a grand beginning. But over the centuries much of Ionian science would be forgotten—only to be rediscovered or reinvented, sometimes more than once.

According to legend, the first mathematical formulation of what we might today call a law of nature dates back to an Ionian named Pythagoras (c. 580 BC–c. 490 BC), famous for the theorem named after him: that the square of the hypotenuse (longest side) of a right triangle equals the sum of the squares of the other two sides. Pythagoras is said to have discovered the numerical relationship between the length of the strings used in musical instruments and the harmonic combinations of the sounds. In today's language we would describe that relationship by saying

Ionia Scholars in ancient Ionia were among the first to explain natural phenomena through laws of nature rather than myth or theology.

that the frequency—the number of vibrations per second—of a string vibrating under fixed tension is inversely proportional to the length of the string. From the practical point of view, this explains why bass guitars must have longer strings than ordinary guitars. Pythagoras probably did not really discover this—he also did not discover the theorem that bears his name—but there is evidence that some relation between string length and pitch was known in his day. If so, one could call that simple mathematical formula the first instance of what we now know as theoretical physics.

Apart from the Pythagorean law of strings, the only physical laws known correctly to the ancients were three laws detailed by Archimedes (c. 287 BC–c. 212 BC), by far the most eminent physicist of antiquity. In today's terminology, the law of the lever explains that small forces can lift large weights because the lever amplifies a force according to the ratio of the distances from the lever's fulcrum. The law of buoyancy states that any object immersed in a fluid will experience an upward force equal to the weight of the displaced fluid. And the law of reflection asserts that the angle between a beam of light and a mirror is equal to the angle between the mirror and the reflected beam. But Archimedes did not call them laws, nor did he explain them with reference to observation and measurement. Instead he treated them as if they were purely mathematical theorems, in an axiomatic system much like the one Euclid created for geometry.

As the Ionian influence spread, there appeared others who saw that the universe possesses an internal order, one that could be understood through observation and reason. Anaximander (c. 610 BC–c. 546 BC), a friend and possibly a student of Thales, argued that since human infants are helpless at birth, if the first human had somehow appeared on earth as an infant, it would not have survived. In what may have been humanity's first inkling of evolution, people, Anaximander reasoned, must therefore have evolved from other animals whose young are hardier. In Sicily, Empedocles (c. 490 BC–c. 430 BC) observed the use of an instrument called a clepsydra. Sometimes used as a ladle, it consisted of a sphere with an open neck and small holes in its bottom. When immersed in water it would fill, and if the open neck was then covered, the clepsydra could be lifted out without the water in it falling through the holes. Empedocles noticed that if you cover the neck before you immerse it, a clepsydra does not fill. He reasoned that something invisible must be preventing the water from entering the sphere through the holes—he had discovered the material substance we call air.

Around the same time Democritus (c. 460 BC–c. 370 BC), from an Ionian colony in northern Greece, pondered what happened when you break or cut an object into pieces. He argued that you ought not to be able to continue the process indefinitely. Instead he postulated that everything, including all living beings, is made of fundamental particles that cannot be cut or broken into parts. He named these ultimate particles atoms, from the Greek adjective meaning "uncuttable". Democritus believed that every material phenomenon is a product of the collision of atoms. In his view, dubbed atomism, all atoms move around in space, and, unless disturbed, move forward indefinitely. Today that idea is called the law of inertia.

The revolutionary idea that we are but ordinary inhabitants of the universe, not special beings distinguished by existing at its centre, was first championed by Aristarchus (c. 310 BC–c. 230 BC), one of the last of the Ionian scientists. Only one of his calculations survives, a complex geometric analysis of careful observations he made of the size of the earth's shadow on the moon during a lunar eclipse. He concluded from his data that the sun must be much larger than the earth. Perhaps inspired by the idea that tiny objects ought to orbit mammoth ones and not the other way around, he became the first person to argue that the earth is not the centre of our planetary system, but rather that it and the other planets orbit the much larger sun. It is a small step from the realization that the earth is just another planet to the idea that our sun is nothing special either. Aristarchus suspected that this was the case and believed that the stars we see in the night sky are actually nothing more than distant suns.

The Ionians were but one of many schools of ancient Greek philosophy, each with different and often contradictory traditions. Unfortunately, the Ionians' view of nature—that it can be explained through general laws and reduced to a simple set of principles—exerted a powerful influence for only a few centuries. One reason is that Ionian theories often seemed to have no place for the notion of free will or purpose, or the concept that gods intervene in the workings of the world. These were startling omissions as profoundly unsettling to many Greek thinkers as they are to many people today. The philosopher Epicurus (341 BC–270 BC), for example, opposed atomism on the grounds that it is "better to follow the myths about the gods than to become a 'slave' to the destiny of natural philosophers". Aristotle too rejected the concept of atoms because he could not accept that human beings were composed of soulless, inanimate objects. The Ionian idea that the universe is not human-centred was a milestone in our understanding of the cosmos, but it was an idea that would be dropped and not picked up again, or commonly accepted, until Galileo, almost twenty centuries later.

As insightful as some of their speculations about nature were, most of the ideas of the ancient Greeks would not pass muster as valid science in modern times. For one, because the Greeks had not invented the scientific method, their theories were not developed with the goal of experimental verification. So if one scholar claimed an atom moved in a straight line until it collided with a second atom and another scholar claimed it moved in a straight line until it bumped into a cyclops, there was no objective way to settle the argument. Also, there was no clear distinction between human and physical laws. In the fifth century BC, for instance, Anaximander wrote that all things arise from a primary substance, and return to it, lest they "pay fine and penalty for their iniquity". And according to the Ionian philosopher Heraclitus (c. 535 BC–c. 475 BC), the sun behaves as it does because otherwise the goddess of justice will hunt it down. Several hundred years later the

Stoics, a school of Greek philosophers that arose around the third century BC, did make a distinction between human statutes and natural laws, but they included rules of human conduct they considered universal—such as veneration of God and obedience to parents—in the category of natural laws. Conversely, they often described physical processes in legal terms and believed them to be in need of enforcement, even though the objects required to "obey" the laws were inanimate. If you think it is hard to get humans to follow traffic laws, imagine convincing an asteroid to move along an ellipse.

This tradition continued to influence the thinkers who succeeded the Greeks for many centuries thereafter. In the thirteenth century the early Christian philosopher Thomas Aquinas (c. 1225–1274) adopted this view and used it to argue for the existence of God, writing, "It is clear that [inanimate bodies] reach their end not by chance but by intention. . . . There is therefore, an intelligent personal being by whom everything in nature is ordered to its end." Even as late as the sixteenth century, the great German astronomer Johannes Kepler (1571–1630) believed that planets had sense perception and consciously followed laws of movement that were grasped by their "mind".

The notion that the laws of nature had to be intentionally obeyed reflects the ancients' focus on *why* nature behaves as it does, rather than on *how* it behaves. Aristotle was one of the leading proponents of that approach, rejecting the idea of science based principally on observation. Precise measurement and mathematical calculation were in any case difficult in ancient times. The base ten number notation we find so convenient for arithmetic dates back only to around AD 700, when the Hindus took the first great strides toward making that subject a powerful tool. The abbreviations for plus and minus didn't come until the fifteenth century. And neither the equal sign nor clocks that could measure times to the second existed before the sixteenth century.

Aristotle, however, did not see problems in measurement and calculation as impediments to developing a physics that could produce quantitative predictions. Rather, he saw no need to make them. Instead, Aristotle built his physics upon principles that appealed to him intellectually. He suppressed facts he found unappealing and focused his efforts on the reasons things happen, with relatively little energy invested in detailing exactly what was happening. Aristotle did adjust his conclusions when their blatant disagreement with observation could not be ignored. But those adjustments were often ad hoc explanations that did little more than paste over the contradiction. In that manner, no matter how severely his theory deviated from actuality, he could always alter it just enough to seem to remove the conflict. For example, his theory of motion specified that heavy bodies fall with a constant speed that is proportional to their weight. To explain the fact that objects clearly pick up speed as they fall, he invented a new principle—that bodies proceed more jubilantly, and hence accelerate, when they come closer to their natural place of rest, a principle that today seems a more apt description of certain people than of inanimate objects. Though Aristotle's theories often had little predictive value, his approach to science dominated Western thought for nearly two thousand years.

The Greeks' Christian successors rejected the idea that the universe is governed by indifferent natural law. They also rejected the idea that humans do not hold a privileged place within that universe. And though the medieval period had no single coherent philosophical system, a common theme was that the universe is God's dollhouse, and religion a far worthier study than the phenomena of nature. Indeed, in 1277 Bishop Tempier of Paris, acting

"If I've learned one thing in my long reign, it's that heat rises."

on the instructions of Pope John XXI, published a list of 219 errors or heresies that were to be condemned. Among the heresies was the idea that nature follows laws, because this conflicts with God's omnipotence. Interestingly, Pope John was killed by the effects of the law of gravity a few months later when the roof of his palace fell in on him.

The modern concept of laws of nature emerged in the seventeenth century. Kepler seems to have been the first scientist to understand the term in the sense of modern science, though as we've said, he retained an animistic view of physical objects. Galileo (1564–1642) did not use the term "law" in his most scientific works (though it appears in some translations of those works). Whether or not he used the word, however, Galileo did uncover a great many laws, and advocated the important principles that observation is the basis of science and that the purpose of science is to research the quantitative relationships that exist between physical phenomena. But the person who first explicitly and rigorously formulated the concept of laws of nature as we understand them was René Descartes (1596–1650).

Descartes believed that all physical phenomena must be explained in terms of the collisions of moving masses, which were governed by three laws—precursors of Newton's famous laws of motion. He asserted that those laws of nature were valid in all places and at all times, and stated explicitly that obedience to these laws does not imply that these moving bodies have minds. Descartes also understood the importance of what we today call "initial conditions". Those describe the state of a system at the beginning of whatever interval of time over which one seeks to make predictions. With a given set of initial conditions, the laws of nature determine how a system will evolve over time, but without a specific set of initial conditions, the evolution cannot be specified. If, for example, at time zero a pigeon directly overhead lets something go, the path of that falling object is determined by Newton's laws. But the outcome will be very different depending on whether, at time zero, the pigeon is sitting still on a telephone wire or flying by at 20 miles per hour. In order to apply the laws of physics one must know how a system started off, or at least its state at some definite time. (One can also use the laws to follow a system backward in time.)

With this renewed belief in the existence of laws of nature came new attempts to reconcile those laws with the concept of God. According to Descartes, God could at will alter the truth or falsity of ethical propositions or mathematical theorems, but not nature. He believed that God ordained the laws of nature but had no choice in the laws; rather, he picked them because the laws we experience are the only possible laws. This would seem to impinge on God's authority, but Descartes got around that by arguing that the laws are unalterable because they are a reflection of God's own intrinsic nature. If that were true, one might think that God still had the choice of creating a variety of different worlds, each corresponding to a different set of initial conditions, but Descartes also denied this. No matter what the arrangement of matter at the beginning of the universe, he argued, over time a world identical to ours would evolve. Moreover, Descartes felt, once God set the world going, he left it entirely alone.

A similar position (with some exceptions) was adopted by Isaac Newton (1643–1727). Newton was the person who won widespread acceptance of the modern concept of a scientific law with his three laws of motion and his law of gravity, which accounted for the orbits of the earth, moon and planets, and explained phenomena such as the tides. The handful of equations he created, and the elaborate mathematical framework we have since derived from them, are still taught today, and employed whenever an architect designs a building, an engineer designs a car, or a physicist calculates how to aim a rocket meant to land on Mars. As the poet Alexander Pope said:

> *Nature and Nature's laws lay hid in night:*
> *God said,* Let Newton be! *and all was light.*

Today most scientists would say a law of nature is a rule that is based upon an observed regularity and provides predictions that go beyond the immediate situations upon which it is based. For example, we might notice that the sun has risen in the east every morning of our lives, and postulate the law, "The sun always rises in the east." This is a generalization that goes beyond our limited observations of the rising sun and makes testable predictions about the future. On the other hand, a statement such as, "The computers in this office are black" is not a law of nature because it relates only to the computers within the office and makes no predictions such as, "If my office purchases a new computer, it will be black."

Our modern understanding of the term "law of nature" is an issue philosophers argue at length, and it is a more subtle question than one may at first think. For example, the philosopher John W. Carroll compared the statement, "All gold spheres are less than a mile in diameter" to a statement like, "All uranium-235 spheres are less than a mile in diameter." Our observations of the world tell us that there are no gold spheres larger than a mile wide, and we can be pretty confident there never will be. Still, we have no reason to believe that there couldn't be one, and so the statement is not considered a law. On the other hand, the statement, "All uranium-235 spheres are less than a mile in diameter" could be thought of as a law of nature because, according to what we know about nuclear physics, once a sphere of uranium-235 grew to a diameter greater than about six inches, it would demolish itself in a nuclear explosion. Hence we can be sure that such spheres do not exist. (Nor would it be a good idea to try to make one!) This distinction matters because it illustrates that not all generalizations we observe can be thought of as laws of nature, and that most laws of nature exist as part of a larger, interconnected system of laws.

In modern science laws of nature are usually phrased in mathematics. They can be either exact or approximate, but they must have been observed to hold without exception—if not universally, then at least under a stipulated set of conditions. For example, we now know that Newton's laws must be modified if objects are moving at velocities near the speed of light. Yet we still consider

Newton's laws to be laws because they hold, at least to a very good approximation, for the conditions of the everyday world, in which the speeds we encounter are far below the speed of light.

If nature is governed by laws, three questions arise:

1. What is the origin of the laws?
2. Are there any exceptions to the laws, i.e., miracles?
3. Is there only one set of possible laws?

These important questions have been addressed in varying ways by scientists, philosophers and theologians. The answer traditionally given to the first question—the answer of Kepler, Galileo, Descartes and Newton—was that the laws were the work of God. However, this is no more than a definition of God as the embodiment of the laws of nature. Unless one endows God with some other attributes, such as being the God of the Old Testament, employing God as a response to the first question merely substitutes one mystery for another. So if we involve God in the answer to the first question, the real crunch comes with the second question: Are there miracles, exceptions to the laws?

Opinions about the answer to the second question have been sharply divided. Plato and Aristotle, the most influential ancient Greek writers, held that there can be no exceptions to the laws. But if one takes the biblical view, then God not only created the laws but can be appealed to by prayer to make exceptions—to heal the terminally ill, to bring premature ends to droughts, or to reinstate croquet as an Olympic sport. In opposition to Descartes' view, almost all Christian thinkers maintained that God must be able to suspend the laws to accomplish miracles. Even Newton believed in miracles of a sort. He thought that the orbit of the planets would be unstable because the gravitational attraction of one planet for another would cause disturbances to the orbits that would grow with time and would result in the planets either falling into the sun or being flung out of the solar system. God must keep on resetting the orbits, he believed, or "wind the celestial watch, lest it run down". However, Pierre-Simon, marquis de Laplace (1749–1827), commonly known as Laplace, argued that the perturbations would be periodic, that is, marked by repeated cycles, rather than being cumulative. The solar system would thus reset itself, and there would be no need for divine intervention to explain why it had survived to the present day.

It is Laplace who is usually credited with first clearly postulating scientific determinism: given the state of the universe at one time, a complete set of laws fully determines both the future and the past. This would exclude the possibility of miracles or an active role for God. The scientific determinism that Laplace formulated is the modern scientist's answer to question two. It is, in fact, the basis of all modern science, and a principle that is important throughout this book. A scientific law is not a scientific law if it holds only when some supernatural being decides not to intervene. Recognizing this, Napoleon is said to have asked Laplace how God fit into this picture. Laplace replied: "Sire, I have not needed that hypothesis."

Since people live in the universe and interact with the other objects in it, scientific determinism must hold for people as well. Many, however, while accepting that scientific determinism governs physical processes, would make an exception for human behaviour because they believe we have free will. Descartes, for instance, in order to preserve the idea of free will, asserted that the human mind was something different from the physical world and did not follow its laws. In his view a person consists of two ingredients, a body and a soul. Bodies are nothing but ordinary

"I think you should be more explicit here in step two."

machines, but the soul is not subject to scientific law. Descartes was very interested in anatomy and physiology and regarded a tiny organ in the centre of the brain, called the pineal gland, as the principal seat of the soul. That gland, he believed, was the place where all our thoughts are formed, the wellspring of our free will.

Do people have free will? If we have free will, where in the evolutionary tree did it develop? Do blue-green algae or bacteria have free will, or is their behaviour automatic and within the realm of scientific law? Is it only multicelled organisms that have free will, or only mammals? We might think that a chimpanzee is exercising free will when it chooses to chomp on a banana, or a cat when it rips up your sofa, but what about the roundworm called *Caenorhabditis elegans*—a simple creature made of only 959 cells? It probably never thinks, "That was damn tasty bacteria I got to dine on back there," yet it too has a definite preference in food and will either settle for an unattractive meal or go foraging for something better, depending on recent experience. Is that the exercise of free will?

Though we feel that we can choose what we do, our

understanding of the molecular basis of biology shows that biological processes are governed by the laws of physics and chemistry and therefore are as determined as the orbits of the planets. Recent experiments in neuroscience support the view that it is our physical brain, following the known laws of science, that determines our actions, and not some agency that exists outside those laws. For example, a study of patients undergoing awake brain surgery found that by electrically stimulating the appropriate regions of the brain, one could create in the patient the desire to move the hand, arm or foot, or to move the lips and talk. It is hard to imagine how free will can operate if our behaviour is determined by physical law, so it seems that we are no more than biological machines and that free will is just an illusion.

While conceding that human behaviour is indeed determined by the laws of nature, it also seems reasonable to conclude that the outcome is determined in such a complicated way and with so many variables as to make it impossible in practice to predict. For that one would need a knowledge of the initial state of each of the thousand trillion trillion molecules in the human body and to solve something like that number of equations. That would take a few billion years, which would be a bit late to duck when the person opposite aimed a blow.

Because it is so impractical to use the underlying physical laws to predict human behaviour, we adopt what is called an effective theory. In physics, an effective theory is a framework created to model certain observed phenomena without describing in detail all of the underlying processes. For example, we cannot solve exactly the equations governing the gravitational interactions of every atom in a person's body with every atom in the earth. But for all practical purposes the gravitational force between a person and the earth can be described in terms of just a few numbers, such as the person's total mass. Similarly, we cannot solve the equations governing the behaviour of complex atoms and molecules, but we have developed an effective theory called chemistry that provides an adequate explanation of how atoms and molecules behave in chemical reactions without accounting for every detail of the interactions. In the case of people, since we cannot solve the equations that determine our behaviour, we use the effective theory that people have free will. The study of our will, and of the behaviour that arises from it, is the science of psychology. Economics is also an effective theory, based on the notion of free will plus the assumption that people evaluate their possible alternative courses of action and choose the best. That effective theory is only moderately successful in predicting behaviour because, as we all know, decisions are often not rational or are based on a defective analysis of the consequences of the choice. That is why the world is in such a mess.

The third question addresses the issue of whether the laws that determine both the universe and human behaviour are unique. If your answer to the first question is that God created the laws, then this question asks, did God have any latitude in choosing them? Both Aristotle and Plato believed, like Descartes and later Einstein, that the principles of nature exist out of "necessity", that is, because they are the only rules that make logical sense. Due to his belief in the origin of the laws of nature in logic, Aristotle and his followers felt that one could "derive" those laws without paying a lot of attention to how nature actually behaved. That, and the focus on why objects follow rules rather than on the specifics of what the rules are, led him to mainly qualitative laws that were often wrong and in any case did not prove very useful, even if they did dominate scientific thought for many centuries. It was only much later that people such as Galileo dared to challenge the authority of Aristotle and observe what nature actually did, rather than what pure "reason" said it ought to do.

This book is rooted in the concept of scientific determinism, which implies that the answer to question two is that there are no miracles, or exceptions to the laws of nature. We will, however, return to address in depth questions one and three, the issues of how the laws arose and whether they are the only possible laws. But first, in the next chapter, we will address the issue of what it is that the laws of nature describe. Most scientists would say they are the mathematical reflection of an external reality that exists independent of the observer who sees it. But as we ponder the manner in which we observe and form concepts about our surroundings, we bump into the question, do we really have reason to believe that an objective reality exists?

UNITED STATES
20c

3

WHAT IS REALITY?

A FEW YEARS AGO the city council of Monza, Italy, barred pet owners from keeping goldfish in curved goldfish bowls. The measure's sponsor explained the measure in part by saying that it is cruel to keep a fish in a bowl with curved sides because, gazing out, the fish would have a distorted view of reality. But how do we know we have the true, undistorted picture of reality? Might not we ourselves also be inside some big goldfish bowl and have our vision distorted by an enormous lens? The goldfish's picture of reality is different from ours, but can we be sure it is less real?

The goldfish view is not the same as our own, but goldfish could still formulate scientific laws governing the motion of the objects they observe outside their bowl. For example, due to the distortion, a freely moving object that we would observe to move in a straight line would be observed by the goldfish to move along a curved path. Nevertheless, the goldfish could formulate scientific laws from their distorted frame of reference that would always hold true and that would enable them to make predictions about the future motion of objects outside the bowl. Their laws would be more complicated than the laws in our frame, but simplicity is a matter of taste. If a goldfish formulated such a theory, we would have to admit the goldfish's view as a valid picture of reality.

A famous example of different pictures of reality is the model introduced around AD 150 by Ptolemy (c. 85—c. 165) to describe the motion of the celestial bodies. Ptolemy published his work in a thirteen-book treatise usually known under its Arabic title,

Almagest. The *Almagest* begins by explaining reasons for thinking that the earth is spherical, motionless, positioned at the centre of the universe, and negligibly small in comparison to the distance of the heavens. Despite Aristarchus's heliocentric model, these beliefs had been held by most educated Greeks at least since the time of Aristotle, who believed for mystical reasons that the earth should be at the centre of the universe. In Ptolemy's model the earth stood still at the centre and the planets and the stars moved around it in complicated orbits involving epicycles, like wheels on wheels.

The Ptolemaic Universe In Ptolemy's view, we lived at the centre of the universe.

This model seemed natural because we don't feel the earth under our feet moving (except in earthquakes or moments of passion). Later European learning was based on the Greek sources that had been passed down, so that the ideas of Aristotle and Ptolemy became the basis for much of Western thought. Ptolemy's model of the cosmos was adopted by the Catholic Church and held as official doctrine for fourteen hundred years. It was not until 1543 that an alternative model was put forward by Copernicus in his book *De revolutionibus orbium coelestium* (*On the Revolutions of the Celestial Spheres*), published only in the year of his death (though he had worked on his theory for several decades).

Copernicus, like Aristarchus some seventeen centuries earlier, described a world in which the sun was at rest and the planets revolved around it in circular orbits. Though the idea wasn't new, its revival was met with passionate resistance. The Copernican model was held to contradict the Bible, which was interpreted as saying that the planets moved around the earth, even though the Bible never clearly stated that. In fact, at the time the Bible was written people believed the earth was flat. The Copernican model led to a furious debate as to whether the earth was at rest, culminating in Galileo's trial for heresy in 1633 for advocating the Copernican model, and for thinking "that one may hold and defend as probable an opinion after it has been declared and defined contrary to the Holy Scripture". He was found guilty, confined to house arrest for the rest of his life, and forced to recant. He is said to have muttered under his breath "*Eppur si muove*", "But still it moves". In 1992 the Roman Catholic Church finally acknowledged that it had been wrong to condemn Galileo.

So which is real, the Ptolemaic or Copernican system? Although it is not uncommon for people to say that Copernicus proved Ptolemy wrong, that is not true. As in the case of our normal view versus that of the goldfish, one can use either picture as a model of the universe, for our observations of the heavens can be explained by assuming either the earth or the sun to be at rest. Despite its role in philosophical debates over the nature of our universe, the real advantage of the Copernican system is simply that the equations of motion are much simpler in the frame of reference in which the sun is at rest.

A different kind of alternative reality occurs in the science fiction film *The Matrix*, in which the human race is unknowingly living in a simulated virtual reality created by intelligent computers to keep them pacified and content while the computers suck their bioelectrical energy (whatever that is). Maybe this is not so far-fetched, because many people prefer to spend their time in the simulated reality of websites such as Second Life. How do we know we are not just characters in a computer-generated soap opera? If we lived in a synthetic imaginary world, events would not necessarily have any logic or consistency or obey any laws. The aliens in control might find it more interesting or amusing to see our reactions, for example, if the full moon split in half, or everyone in the world on a diet developed an uncontrollable craving for banana cream pie. But if the aliens did enforce consistent laws, there is no way we could tell there was another reality behind the simulated one. It would be easy to call the world the aliens live in the "real" one and the synthetic world a "false" one. But if—like us—the beings in the simulated world could not gaze into their universe from the outside, there would be no reason for them to doubt their own pictures of reality. This is a modern version of the idea that we are all figments of someone else's dream.

These examples bring us to a conclusion that will be important in this book: *there is no picture- or theory-independent concept of reality.* Instead we will adopt a view that we will call model-dependent

realism: the idea that a physical theory or world picture is a model (generally of a mathematical nature) and a set of rules that connect the elements of the model to observations. This provides a framework with which to interpret modern science.

Philosophers from Plato onward have argued over the years about the nature of reality. Classical science is based on the belief that there exists a real external world whose properties are definite and independent of the observer who perceives them. According to classical science, certain objects exist and have physical properties, such as speed and mass, that have well-defined values. In this view our theories are attempts to describe those objects and their properties, and our measurements and perceptions correspond to them. Both observer and observed are parts of a world that has an objective existence, and any distinction between them has no meaningful significance. In other words, if you see a herd of zebras fighting for a spot in the car park, it is because there really is a herd of zebras fighting for a spot in the car park. All other observers who look will measure the same properties, and the herd will have those properties whether anyone observes them or not. In philosophy that belief is called realism.

Though realism may be a tempting viewpoint, as we'll see later, what we know about modern physics makes it a difficult one to defend. For example, according to the principles of quantum physics, which is an accurate description of nature, a particle has neither a definite position nor a definite velocity unless and until those quantities are measured by an observer. It is therefore *not* correct to say that a measurement gives a certain result because the quantity being measured had that value at the time of the measurement. In fact, in some cases individual objects don't even have an independent existence but rather exist only as part of an ensemble of many. And if a theory called the holographic principle proves correct, we and our four-dimensional world may be shadows on the boundary of a larger, five-dimensional space-time. In that case, our status in the universe is analogous to that of the goldfish.

Strict realists often argue that the proof that scientific theories represent reality lies in their success. But different theories can successfully describe the same phenomenon through disparate conceptual frameworks. In fact, many scientific theories that had proven successful were later replaced by other, equally successful theories based on wholly new concepts of reality.

Traditionally those who didn't accept realism have been called anti-realists. Anti-realists suppose a distinction between empirical knowledge and theoretical knowledge. They typically argue that observation and experiment are meaningful but that theories are no more than useful instruments that do not embody any deeper truths underlying the observed phenomena. Some anti-realists have even wanted to restrict science to things that can be observed. For that reason, many in the nineteenth century rejected the idea of atoms on the grounds that we would never see one. George Berkeley (1685–1753) even went as far as to say that nothing exists except the mind and its ideas. When a friend remarked to English author and lexicographer Dr. Samuel Johnson (1709–1784) that Berkeley's claim could not possibly be refuted, Johnson is said to have responded by walking over to a large stone, kicking it, and proclaiming, "I refute it thus." Of course the pain Dr. Johnson experienced in his foot was also an idea in his mind, so he wasn't really refuting Berkeley's ideas. But his act did illustrate the view of philosopher David Hume (1711–1776), who wrote that although we have no rational grounds for believing in an objective reality, we also have no choice but to act as if it is true.

Model-dependent realism short-circuits all this argument and discussion between the realist and anti-realist schools of thought.

"You both have something in common. Dr. Davis has discovered a particle which nobody has seen, and Prof. Higbe has discovered a galaxy which nobody has seen."

According to model-dependent realism, it is pointless to ask whether a model is real, only whether it agrees with observation. If there are two models that both agree with observation, like the goldfish's picture and ours, then one cannot say that one is more real than another. One can use whichever model is more convenient in the situation under consideration. For example, if one were inside the bowl, the goldfish's picture would be useful, but for those outside, it would be very awkward to describe events from a distant galaxy in the frame of a bowl on earth, especially because the bowl would be moving as the earth orbits the sun and spins on its axis.

We make models in science, but we also make them in everyday life. Model-dependent realism applies not only to scientific models but also to the conscious and subconscious mental models we all create in order to interpret and understand the everyday world. There is no way to remove the observer—us—from our perception of the world, which is created through our sensory processing and through the way we think and reason. Our perception—and hence the observations upon which our theories are based—is not direct, but rather is shaped by a kind of lens, the interpretive structure of our human brains.

Model-dependent realism corresponds to the way we perceive objects. In vision, one's brain receives a series of signals down the optic nerve. Those signals do not constitute the sort of image you would accept on your television. There is a blind spot where the optic nerve attaches to the retina, and the only part of your field of vision with good resolution is a narrow area of about 1 degree of visual angle around the retina's centre, an area the width of your thumb when held at arm's length. And so the raw data sent to the brain are like a badly pixilated picture with a hole in it. Fortunately, the human brain processes that data, combining the input from both eyes, filling in gaps on the assumption that the visual properties of neighbouring locations are similar and interpolating. Moreover, it reads a two-dimensional array of data from the retina and creates from it the impression of three-dimensional space. The brain, in other words, builds a mental picture or model.

The brain is so good at model building that if people are fitted with glasses that turn the images in their eyes upside down, their brains, after a time, change the model so that they again see things the right way up. If the glasses are then removed, they see the world upside down for a while, then again adapt. This shows that what one means when one says "I see a chair" is merely that one has used the light scattered by the chair to build a mental image or model of the chair. If the model is upside down, with luck one's brain will correct it before one tries to sit on the chair.

Another problem that model-dependent realism solves, or at least avoids, is the meaning of existence. How do I know that a table still exists if I go out of the room and can't see it? What does it mean to say that things we can't see, such as electrons or quarks—the particles that are said to make up the proton and neutron—exist? One could have a model in which the table disappears when I leave the room and reappears in the same position when I come back, but that would be awkward, and what if something happened when I was out, like the ceiling falling in? How, under the table-disappears-when-I-leave-the-room model, could I account for the fact that the next time I enter, the table reappears broken, under the debris of the ceiling? The model in which the table stays put is much simpler and agrees with observation. That is all one can ask.

In the case of subatomic particles that we can't see, electrons are a useful model that explains observations like tracks in a cloud chamber and the spots of light on a television tube, as well as many other phenomena. It is said that the electron was discovered in 1897 by British physicist J. J. Thomson at the Cavendish Laboratory at Cambridge University. He was experimenting with currents of electricity inside empty glass tubes, a phenomenon known as cathode rays. His experiments led him to the bold conclusion that the mysterious rays were composed of minuscule "corpuscles" that were material constituents of atoms, which were then thought to be the indivisible fundamental unit of matter. Thomson did not "see" an electron, nor was his speculation directly or unambiguously demonstrated by his experiments. But the model has proved crucial in applications from fundamental science to engineering, and today all physicists believe in electrons, even though you cannot see them.

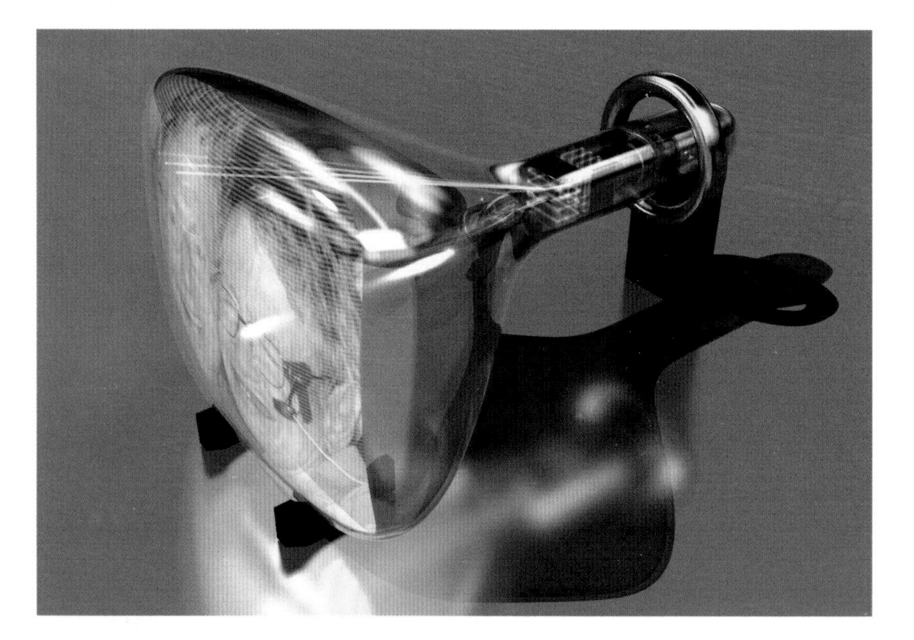

Cathode Rays We can't see individual electrons, but we can see effects they produce.

Quarks, which we also cannot see, are a model to explain the properties of the protons and neutrons in the nucleus of an atom. Though protons and neutrons are said to be made of quarks, we will never observe a quark because the binding force between quarks increases with separation, and hence isolated, free quarks cannot exist in nature. Instead, they always occur in groups of three (protons and neutrons), or in pairings of a quark and an anti-quark (pi mesons), and behave as if they were joined by rubber bands.

The question of whether it makes sense to say quarks really exist if you can never isolate one was a controversial issue in the years after the quark model was first proposed. The idea that certain particles were made of different combinations of a few sub-subnuclear particles provided an organizing principle that yielded a simple and attractive explanation for their properties. But although physicists were accustomed to accepting particles that were only inferred to exist from statistical blips in data pertaining to the scattering of other particles, the idea of assigning reality to a particle that might be, in principle, unobservable was too much for many physicists. Over the years, however, as the quark model led to more and more correct predictions, that opposition faded. It is certainly possible that some alien beings with seventeen arms, infrared eyes and a habit of blowing clotted cream out their ears would make the same experimental observations that we do, but describe them without quarks. Nevertheless, according to model-dependent realism, quarks exist in a model that agrees with our observations of how subnuclear particles behave.

Model-dependent realism can provide a framework to discuss questions such as: if the world was created a finite time ago, what happened before that? An early Christian philosopher, St. Augustine (354–430), said that the answer was not that God was

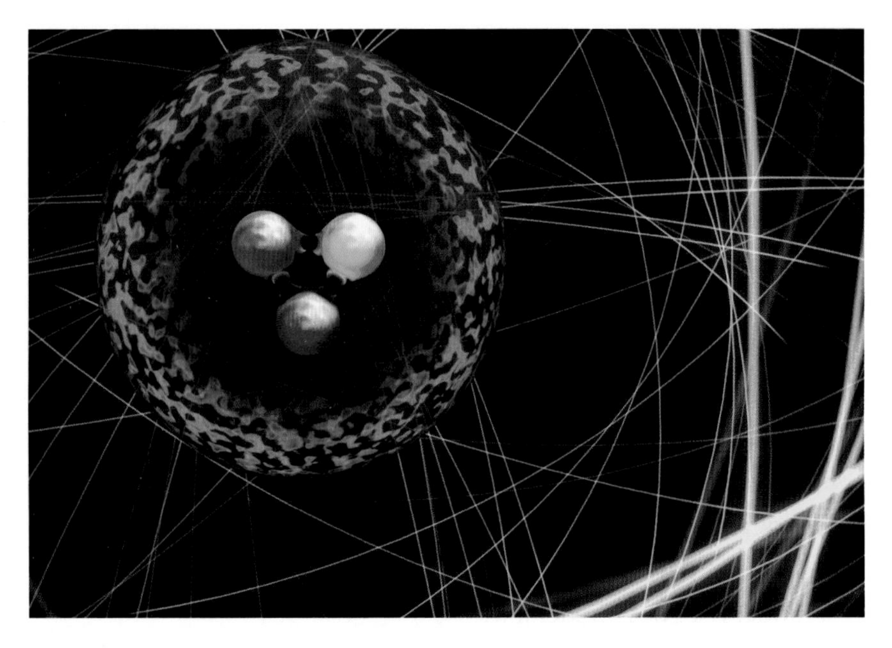

Quarks The concept of quarks is a vital element of our theories of fundamental physics even though individual quarks cannot be observed.

preparing hell for people who ask such questions, but that time was a property of the world that God created and that time did not exist before the creation, which he believed had occurred not that long ago. That is one possible model, which is favoured by those who maintain that the account given in Genesis is literally true even though the world contains fossil and other evidence that makes it look much older. (Were they put there to fool us?) One can also have a different model, in which time continues back 13.7 billion years to the big bang. The model that explains the most about our present observations, including the historical and geological evidence, is the best representation we have of the past. The second model can explain the fossil and radioactive records and the fact that we receive light from galaxies millions of light-years from us, and so this model—the big bang theory—is more useful than the first. Still, neither model can be said to be more real than the other.

Some people support a model in which time goes back even further than the big bang. It is not yet clear whether a model in which time continued back beyond the big bang would be better at explaining present observations because it seems the laws of the evolution of the universe may break down at the big bang. If they do, it would make no sense to create a model that encompasses time before the big bang, because what existed then would have no observable consequences for the present, and so we might as well stick with the idea that the big bang was the creation of the world.

A model is a good model if it:

1. Is elegant
2. Contains few arbitrary or adjustable elements
3. Agrees with and explains all existing observations
4. Makes detailed predictions about future observations that can disprove or falsify the model if they are not borne out.

For example, Aristotle's theory that the world was made of four elements, earth, air, fire and water, and that objects acted to fulfil their purpose was elegant and didn't contain adjustable elements. But in many cases it didn't make definite predictions, and when it did, the predictions weren't always in agreement with observation. One of these predictions was that heavier objects should fall faster because their purpose is to fall. Nobody seemed to have thought that it was important to test this until Galileo.

There is a story that he tested it by dropping weights from the Leaning Tower of Pisa. This is probably apocryphal, but we do know he rolled different weights down an inclined plane and observed that they all gathered speed at the same rate, contrary to Aristotle's prediction.

The above criteria are obviously subjective. Elegance, for example, is not something easily measured, but it is highly prized among scientists because laws of nature are meant to economically compress a number of particular cases into one simple formula. Elegance refers to the form of a theory, but it is closely related to a lack of adjustable elements, since a theory jammed with fudge factors is not very elegant. To paraphrase Einstein, a theory should be as simple as possible, but not simpler. Ptolemy added epicycles to the circular orbits of the heavenly bodies in order that his model might accurately describe their motion. The model could have been made more accurate by adding epicycles to the epicycles, or even epicycles to those. Though added complexity could make the model more accurate, scientists view a model that is contorted to match a specific set of observations as unsatisfying, more of a catalogue of data than a theory likely to embody any useful principle.

We'll see in Chapter 5 that many people view the "standard model", which describes the interactions of the elementary particles of nature, as inelegant. That model is far more successful than Ptolemy's epicycles. It predicted the existence of several new particles before they were observed, and described the outcome of numerous experiments over several decades to great precision. But it contains dozens of adjustable parameters whose values must be fixed to match observations, rather than being determined by the theory itself.

As for the fourth point, scientists are always impressed when new and stunning predictions prove correct. On the other hand, when a model is found lacking, a common reaction is to say the experiment was wrong. If that doesn't prove to be the case, people still often don't abandon the model but instead attempt to save it through modifications. Although physicists are indeed tenacious in their attempts to rescue theories they admire, the tendency to modify a theory fades to the degree that the alterations become artificial or cumbersome, and therefore "inelegant".

If the modifications needed to accommodate new observations become too baroque, it signals the need for a new model. One example of an old model that gave way under the weight of new observations was the idea of a static universe. In the 1920s, most physicists believed that the universe was static, or unchanging in size. Then, in 1929, Edwin Hubble published his observations showing that the universe is expanding. But Hubble did not directly observe the universe expanding. He observed the light emitted by galaxies. That light carries a characteristic signature, or spectrum, based on each galaxy's composition, which changes by a known amount if the galaxy is moving relative to us. Therefore, by analyzing the spectra of distant galaxies, Hubble was able to determine their velocities. He had expected to find as many galaxies moving away from us as moving toward us. Instead he found that nearly all galaxies were moving away from us, and the farther away they were, the faster they were moving. Hubble concluded that the universe is expanding, but others, trying to hold on to the earlier model, attempted to explain his observations within the context of the static universe. For example, Caltech physicist Fritz Zwicky suggested that for some yet unknown reason light might slowly lose energy as it travels great distances. This decrease in energy would correspond to a change in the light's spectrum, which Zwicky suggested could mimic Hubble's observations. For decades after Hubble, many scientists continued to hold on to the steady-state theory. But the most natural model was Hubble's, that of an expanding universe, and it has come to be the accepted one.

In our quest to find the laws that govern the universe we have formulated a number of theories or models, such as the four-element theory, the Ptolemaic model, the phlogiston theory, the big bang theory, and so on. With each theory or model, our concepts of reality and of the fundamental constituents of the universe have changed. For example, consider the theory of light.

Refraction Newton's model of light could explain why light bent when it passed from one medium to another, but it could not explain another phenomenon we now call Newton's rings.

Newton thought that light was made up of little particles or corpuscles. This would explain why light travels in straight lines, and Newton also used it to explain why light is bent or refracted when it passes from one medium to another, such as from air to glass or air to water.

The corpuscle theory could not, however, be used to explain a phenomenon that Newton himself observed, which is known as Newton's rings. Place a lens on a flat reflecting plate and illuminate it with light of a single colour, such as a sodium light. Looking down from above, one will see a series of light and dark rings centred on where the lens touches the surface. This would be difficult to explain with the particle theory of light, but it can be accounted for in the wave theory.

According to the wave theory of light, the light and dark rings are caused by a phenomenon called interference. A wave, such as a water wave, consists of a series of crests and troughs. When waves collide, if those crests and troughs happen to correspond, they reinforce each other, yielding a larger wave. That is called constructive interference. In that case the waves are said to be "in phase". At the other extreme, when the waves meet, the crests of one wave might coincide with the troughs of the other. In that case the waves cancel each other and are said to be "out of phase". That situation is called destructive interference.

In Newton's rings the bright rings are located at distances from the centre where the separation between the lens and the reflecting plate is such that the wave reflected from the lens differs from the wave reflected from the plate by an integral (1, 2, 3, . . .) number of wavelengths, creating constructive interference. (A wavelength is the distance between one crest or trough of a wave and the next.) The dark rings, on the other hand, are located at distances from the

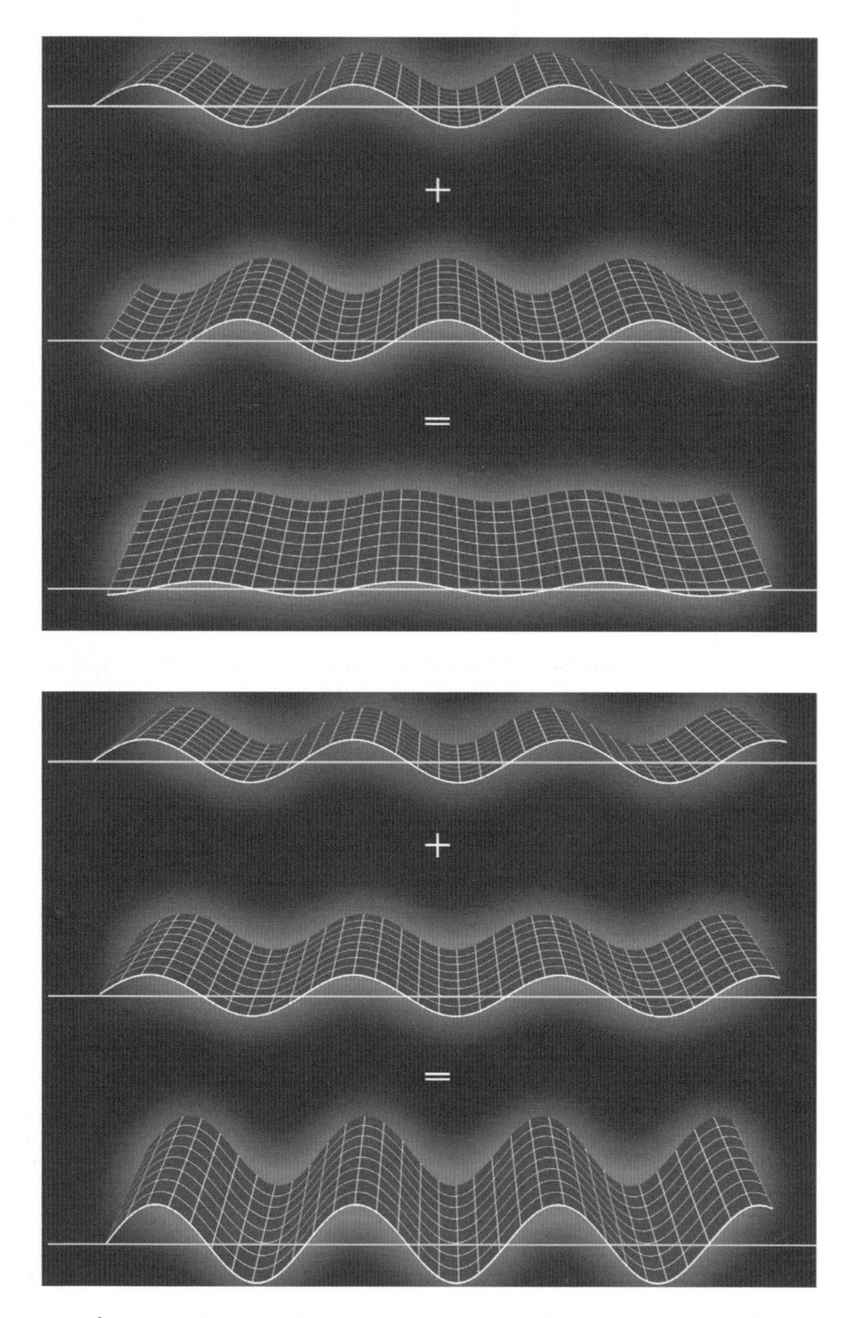

Interference Like people, when waves meet they can tend to either enhance or diminish each other.

centre where the separation between the two reflected waves is a half-integral (½, 1½, 2½, . . .) number of wavelengths, causing destructive interference—the wave reflected from the lens cancels the wave reflected from the plate.

In the nineteenth century, this was taken as confirming the wave theory of light and showing that the particle theory was wrong. However, early in the twentieth century Einstein showed that the photoelectric effect (now used in television and digital cameras) could be explained by a particle or quantum of light striking an atom and knocking out an electron. Thus light behaves as both particle and wave.

The concept of waves probably entered human thought because people watched the ocean, or a puddle after a pebble fell

Puddle Interference The concept of interference shows up in everyday life in bodies of water, from puddles to oceans.

into it. In fact, if you have ever dropped two pebbles into a puddle, you have probably seen interference at work, as in the picture above. Other liquids were observed to behave in a similar fashion, except perhaps wine if you've had too much. The idea of particles was familiar from rocks, pebbles and sand. But this wave/particle duality—the idea that an object could be described as either a particle or a wave—is as foreign to everyday experience as is the idea that you can drink a chunk of sandstone.

Dualities like this—situations in which two very different theories accurately describe the same phenomenon—are consistent with model-dependent realism. Each theory can describe and explain certain properties, and neither theory can be said to be better or more real than the other. Regarding the laws that govern the universe, what we can say is this: there seems to be no single mathematical model or theory that can describe every aspect of the universe. Instead, as mentioned in the opening chapter, there seems to be the network of theories called M-theory. Each theory in the M-theory network is good at describing phenomena within a certain range. Wherever their ranges overlap, the various theories in the network agree, so they can all be said to be parts of the same theory. But no single theory within the network can describe every aspect of the universe—all the forces of nature, the particles that feel those forces, and the framework of space and time in which it all plays out. Though this situation does not fulfil the traditional physicists' dream of a single unified theory, it is acceptable within the framework of model-dependent realism.

We will discuss duality and M-theory further in Chapter 5, but before that we turn to a fundamental principle upon which our modern view of nature is based: quantum theory, and in particular, the approach to quantum theory called alternative histories. In that view, the universe does not have just a single existence or history, but rather every possible version of the universe exists simultaneously in what is called a quantum superposition. That may sound as outrageous as the theory in which the table disappears whenever we leave the room, but in this case the theory has passed every experimental test to which it has ever been subjected.

4

ALTERNATIVE HISTORIES

IN 1999 A TEAM OF PHYSICISTS in Austria fired a series of football-shaped molecules towards a barrier. Those molecules, each made of sixty carbon atoms, are sometimes called buckyballs because the architect Buckminster Fuller built buildings of that shape. Fuller's geodesic domes were probably the largest football-shaped objects in existence. The buckyballs were the smallest. The barrier toward which the scientists took their aim had, in effect, two slits through which the buckyballs could pass. Beyond the wall, the physicists situated the equivalent of a screen to detect and count the emergent molecules.

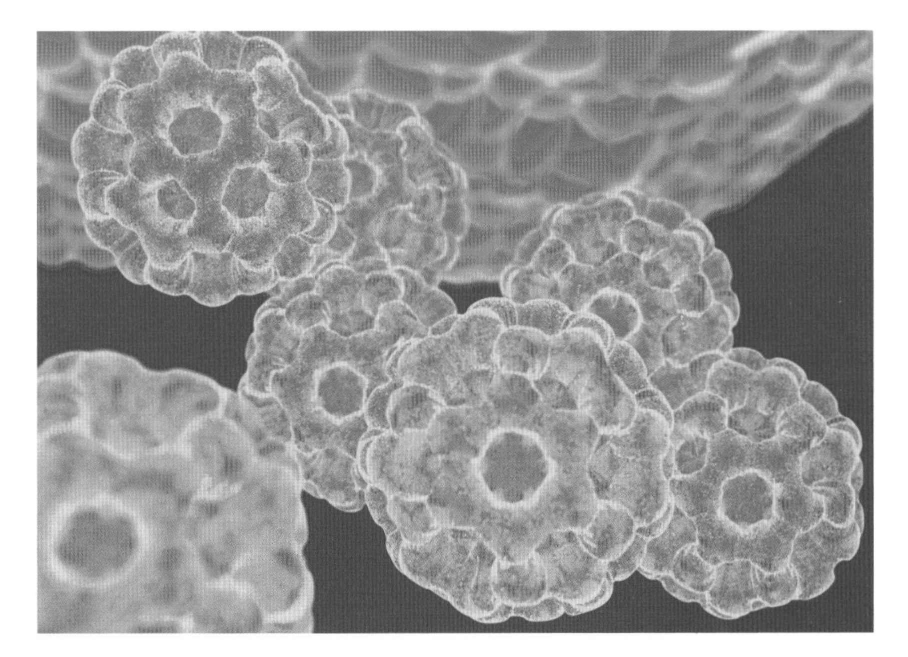

Buckyballs Buckyballs are like microscopic footballs made of carbon atoms.

If we were to set up an analogous experiment with real footballs, we would need a player with somewhat shaky aim but with the ability to launch the balls consistently at a speed of our choosing. We would position this player before a wall in which there are two gaps. On the far side of the wall, and parallel to it, we would place a very long net. Most of the player's shots would hit the wall and bounce back, but some would go through one gap or the other, and into the net. If the gaps were only slightly larger than the balls, two highly collimated streams would emerge on the other side. If the gaps were a bit wider than that, each stream would fan out a little, as shown in the figure below.

Notice that if we closed off one of the gaps, the corresponding stream of balls would no longer get through, but this would have

Two-Slit Football A football player kicking balls at slits in a wall would produce an obvious pattern.

no effect on the other stream. If we reopened the second gap, that would only increase the number of balls that land at any given point on the other side, for we would then get all the balls that passed through the gap that had remained open, plus other balls coming from the newly opened gap. What we observe with both gaps open, in other words, is the sum of what we observe with each gap in the wall separately opened. That is the reality we are accustomed to in everyday life. But that's not what the Austrian researchers found when they fired their molecules.

In the Austrian experiment, opening the second gap did indeed increase the number of molecules arriving at some points on the screen—but it decreased the number at others, as in the figure below. In fact, there were spots where no buckyballs landed when

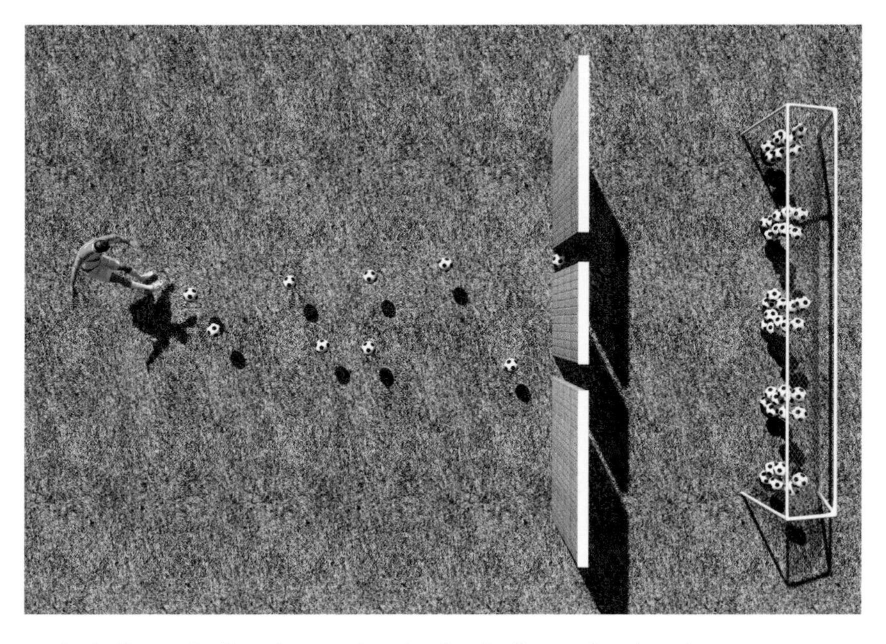

Buckyball Football When molecular footballs are fired at slits in a screen, the resulting pattern reflects unfamiliar quantum laws.

both slits were open but where balls did land when only one or the other gap was open. That seems very odd. How can opening a second gap cause fewer molecules to arrive at certain points?

We can get a clue to the answer by examining the details. In the experiment, many of the molecular footballs landed at a spot centred halfway between where you would expect them to land if the balls went through either one gap or the other. A little farther out from that central position very few molecules arrived, but a bit farther away from the centre than that, molecules were again observed to arrive. This pattern is not the sum of the patterns formed when each gap is opened separately, but you may recognize it from Chapter 3 as the pattern characteristic of interfering waves. The areas where no molecules arrive correspond to regions in which waves emitted from the two gaps arrive out of phase, and create destructive interference; the areas where many molecules arrive correspond to regions where the waves arrive in phase, and create constructive interference.

In the first two thousand or so years of scientific thought, ordinary experience and intuition were the basis for theoretical explanation. As we improved our technology and expanded the range of phenomena that we could observe, we began to find nature behaving in ways that were less and less in line with our everyday experience and hence with our intuition, as evidenced by the experiment with buckyballs. That experiment is typical of the type of phenomena that cannot be encompassed by classical science but are described by what is called quantum physics. In fact, Richard Feynman wrote that the double-slit experiment like the one we described above "contains all the mystery of quantum mechanics".

The principles of quantum physics were developed in the first few decades of the twentieth century after Newtonian theory was found to be inadequate for the description of nature on the atomic—or subatomic—level. The fundamental theories of physics describe the forces of nature and how objects react to them. Classical theories such as Newton's are built upon a framework reflecting everyday experience, in which material objects have an individual existence, can be located at definite locations, follow definite paths, and so on. Quantum physics provides a framework for understanding how nature operates on atomic and subatomic scales, but as we'll see in more detail later, it dictates a completely different conceptual schema, one in which an object's position, path, and even its past and future are not precisely determined. Quantum theories of forces such as gravity or the electromagnetic force are built within that framework.

Can theories built upon a framework so foreign to everyday experience also explain the events of ordinary experience that were modelled so accurately by classical physics? They can, for we and our surroundings are composite structures, made of an unimaginably large number of atoms, more atoms than there are stars in the observable universe. And though the component atoms obey the principles of quantum physics, one can show that the large assemblages that form soccer balls, turnips and jumbo jets—and us—will indeed manage to avoid diffracting through slits. So though the components of everyday objects obey quantum physics, Newton's laws form an effective theory that describes very accurately how the composite structures that form our everyday world behave.

That might sound strange, but there are many instances in science in which a large assemblage appears to behave in a manner that is different from the behaviour of its individual components. The responses of a single neuron hardly portend those of the human brain, nor does knowing about a water molecule tell you much about the behaviour of a lake. In the case of quantum physics, physicists are still working to figure out the details of how Newton's laws emerge from the quantum domain. What we do know is that the components of all objects obey the laws of quantum physics, and the Newtonian laws are a good approximation for describing the way macroscopic objects made of those quantum components behave.

The predictions of Newtonian theory therefore match the view of reality we all develop as we experience the world around us. But individual atoms and molecules operate in a manner profoundly different from that of our everyday experience. Quantum physics is a new model of reality that gives us a picture of the universe. It is a picture in which many concepts fundamental to our intuitive understanding of reality no longer have meaning.

The double-slit experiment was first carried out in 1927 by Clinton Davisson and Lester Germer, experimental physicists at Bell Labs who were studying how a beam of electrons—objects much simpler than buckyballs—interacts with a crystal made of nickel. The fact that matter particles such as electrons behave like water waves was the type of startling experiment that inspired quantum physics. Since this behaviour is not observed on a macroscopic scale, scientists have long wondered just how large and complex something could be and still exhibit such wavelike properties. It would cause quite a stir if the effect could be demonstrated using people or a hippopotamus, but as we've said, in general, the larger the object the less apparent and robust are the quantum effects. So it is unlikely that any zoo animals will be passing wavelike through the bars of their cages. Still, experimental physicists have observed the wave phenomenon with particles of ever-increasing size. Scientists hope to replicate the buckyball experiment someday using a virus, which is not only far bigger but also considered by some to be a living thing.

There are only a few aspects of quantum physics needed to understand the arguments we will make in later chapters. One of the key features is wave/particle duality. That matter particles behave like a wave surprised everyone. That light behaves like a wave no longer surprises anyone. The wavelike behaviour of light seems natural to us and has been considered an accepted fact for almost two centuries. If you shine a beam of light on the two slits in the above experiment, two waves will emerge and meet on the screen. At some points their crests or troughs will coincide and form a bright spot; at others the crests of one beam will meet the troughs of the other, cancelling them, and leaving a dark area. The English physicist Thomas Young performed this experiment in the early nine-

Young's Experiment The buckyball pattern was familiar from the wave theory of light.

teenth century, convincing people that light was a wave and not, as Newton had believed, composed of particles.

Though one might conclude that Newton was wrong to say that light was not a wave, he was right when he said that light can act as if it is composed of particles. Today we call them photons. Just as we are composed of a large number of atoms, the light we see in everyday life is composite in the sense that it is made of a great many photons—even a 1-watt night-light emits a billion billion each second. Single photons are not usually evident, but in the laboratory we can produce a beam of light so faint that it consists of a stream of single photons, which we can detect as individuals just as we can detect individual electrons or buckyballs. And we can repeat Young's experiment employing a beam sufficiently sparse that the photons reach the barrier one at a time, with a few seconds between each arrival. If we do that, and then add up all the individual impacts recorded by the screen on the far side of the barrier, we find that together they build up the same interference pattern that would be built up if we performed the Davisson-Germer experiment but fired the electrons (or buckyballs) at the screen one at a time. To physicists, that was a startling revelation: if individual particles interfere with themselves, then the wave nature of light is the property not just of a beam or of a large collection of photons but of the individual particles.

Another of the main tenets of quantum physics is the uncertainty principle, formulated by Werner Heisenberg in 1926. The uncertainty principle tells us that there are limits to our ability to simultaneously measure certain data, such as the position and velocity of a particle. According to the uncertainty principle, for example, if you multiply the uncertainty in the position of a particle by the uncertainty in its momentum (its mass times its velocity) the result can never be smaller than a certain fixed quantity, called Planck's

"If this is correct, then everything we thought was a wave is really a particle, and everything we thought was a particle is really a wave."

constant. That's a tongue-twister, but its gist can be stated simply: the more precisely you measure speed, the less precisely you can measure position, and vice versa. For instance, if you halve the uncertainty in position, you have to double the uncertainty in velocity. It is also important to note that, compared with everyday units of measurement such as metres, kilograms and seconds, Planck's constant is very small. In fact, if reported in those units, it has the value of about 6/10,000,000,000,000,000,000,000,000,000,000,000. As a result, if you pinpoint a macroscopic object such as a football, with a mass of one-third of a kilogram, to within 1 millimetre in any direction, we can still measure its velocity with a precision far greater than even a billionth of a billionth of a billionth of a kilometre per hour. That's because, measured in these units, the football has a mass of 1/3, and the uncertainty in position is 1/1,000. Neither is enough to account for all those zeroes in

Planck's constant, and so that role falls to the uncertainty in velocity. But in the same units an electron has a mass of .000000000000000000000000000001, so for electrons the situation is quite different. If we measure the position of an electron to a precision corresponding to roughly the size of an atom, the uncertainty principle dictates that we cannot know the electron's speed more precisely than about plus or minus 1,000 kilometres per second, which is not very precise at all.

According to quantum physics, no matter how much information we obtain or how powerful our computing abilities, the outcomes of physical processes cannot be predicted with certainty because they are not *determined* with certainty. Instead, given the initial state of a system, nature determines its future state through a process that is fundamentally uncertain. In other words, nature does not dictate the outcome of any process or experiment, even in the simplest of situations. Rather, it allows a number of different eventualities, each with a certain likelihood of being realized. It is, to paraphrase Einstein, as if God throws the dice before deciding the result of every physical process. That idea bothered Einstein, and so even though he was one of the fathers of quantum physics, he later became critical of it.

Quantum physics might seem to undermine the idea that nature is governed by laws, but that is not the case. Instead it leads us to accept a new form of determinism: given the state of a system at some time, the laws of nature determine the *probabilities* of various futures and pasts rather than determining the future and past with certainty. Though that is distasteful to some, scientists must accept theories that agree with experiment, not their own preconceived notions.

What science does demand of a theory is that it be testable. If the probabilistic nature of the predictions of quantum physics meant it was impossible to confirm those predictions, then quantum theories would not qualify as valid theories. But despite the probabilistic nature of their predictions, we can still test quantum theories. For instance, we can repeat an experiment many times and confirm that the frequency of various outcomes conforms to the probabilities predicted. Consider the buckyball experiment. Quantum physics tells us that nothing is ever located at a definite point because if it were, the uncertainty in momentum would have to be infinite. In fact, according to quantum physics, each particle has some probability of being found anywhere in the universe. So even if the chances of finding a given electron within the double-slit apparatus are very high, there will always be some chance that it could be found instead on the far side of the star Alpha Centauri, or in the shepherd's pie at your office cafeteria. As a result, if you kick a quantum buckyball and let it fly, no amount of skill or knowledge will allow you to say in advance exactly where it will land. But if you repeat that experiment many times, the data you obtain will reflect the probability of finding the ball at various locations, and experimenters have confirmed that the results of such experiments agree with the theory's predictions.

It is important to realize that probabilities in quantum physics are not like probabilities in Newtonian physics, or in everyday life. We can understand this by comparing the patterns built up by the steady stream of buckyballs fired at a screen to the pattern of holes built up by players aiming for the bull's-eye on a dartboard. Unless the players have consumed too much beer, the chances of a dart landing near the centre are greatest, and diminish as you go farther out. As with the buckyballs, any given dart can land anywhere, and over time a pattern of holes that reflects the underlying probabilities will emerge. In everyday life we might reflect that situation by saying that a dart has a certain probability of landing in various spots; but if we say that, unlike the case of the buckyballs, it is only because our knowledge of the conditions of its launch is incomplete. We could improve our description if we knew exactly the manner in which the player released the dart, its angle, spin, velocity, and so forth. In principle, then, we could predict where the dart will land with a precision as great as we desire. Our use of probabilistic terms to describe the outcome of events in everyday life is therefore a reflection not of the intrinsic nature of the process but only of our ignorance of certain aspects of it.

Probabilities in quantum theories are different. They reflect a fundamental randomness in nature. The quantum model of nature encompasses principles that contradict not only our everyday experience but our intuitive concept of reality. Those who find those principles weird or difficult to believe are in good company, the company of great physicists such as Einstein and even Feynman, whose description of quantum theory we will soon present. In fact, Feynman once wrote, "I think I can safely say that nobody understands quantum mechanics." But quantum physics agrees with observation. It has never failed a test, and it has been tested more than any other theory in science.

In the 1940s Richard Feynman had a startling insight regarding the difference between the quantum and Newtonian worlds. Feynman was intrigued by the question of how the interference pattern in the double-slit experiment arises. Recall that the pattern we find when we fire molecules with both slits open is not the sum of the patterns we find when we run the experiment twice, once with just one slit open, and once with only the other open. Instead, when both slits are open we find a series of light and dark bands, the latter being regions in which no particles land. That means that particles that would have landed in the area of the dark band if, say, only slit one was open, do not land there when slit two is also open. It seems as if, somewhere on their journey from source to screen, the particles acquire information about both slits. That kind of behaviour is drastically different from the way things seem to behave in everyday life, in which a ball would follow a path through one of the slits and be unaffected by the situation at the other.

According to Newtonian physics—and to the way the experiment would work if we did it with soccer balls instead of molecules—each particle follows a single well-defined route from its source to the screen. There is no room in this picture for a detour in which the particle visits the neighbourhood of each slit along the way. According to the quantum model, however, the particle is said to have no definite position during the time it is between the starting point and the endpoint. Feynman realized one does not have to interpret that to mean that particles take *no* path as they travel between source and screen. It could mean instead that particles take *every* possible path connecting those points. This, Feynman asserted, is what makes quantum physics different from Newtonian physics. The situation at both slits matters because, rather than following a single definite path, particles take every path, and they take them all *simultaneously*! That sounds like science fiction, but it isn't. Feynman formulated a mathematical expression—the Feynman sum over histories—that reflects this idea and reproduces all the laws of quantum physics. In Feynman's theory the mathematics and physical picture are different from that of the original formulation of quantum physics, but the predictions are the same.

In the double-slit experiment Feynman's ideas mean the particles take paths that go through only one slit or only the other; paths that thread through the first slit, back out through the second slit, and then through the first again; paths that visit the restaurant that serves that great curried shrimp, and then circle Jupiter a few times before heading home; even paths that go across the universe and back. This, in Feynman's view, explains how the particle acquires the information about which slits are open—if a slit is open, the particle takes paths through it. When both slits are open, the paths in which the particle travels through one slit can interfere with the paths in which it travels through the other, causing the interference. It might sound nutty, but for the purposes of most fundamental physics done today—and for the purposes of this book—Feynman's formulation has proved more useful than the original one.

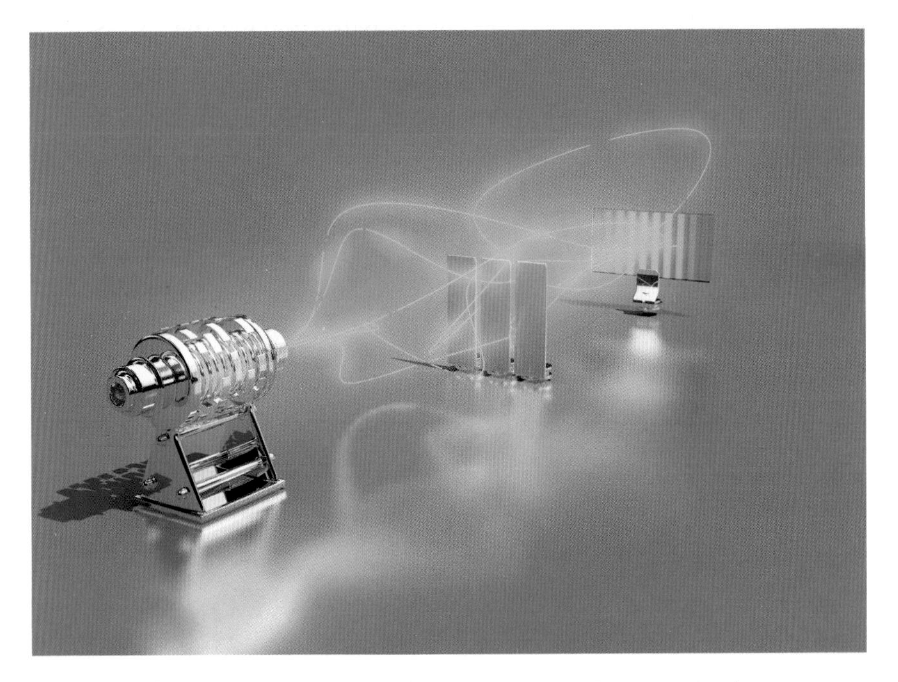

Particle Paths Feynman's formulation of quantum theory provides a picture of why particles such as buckyballs and electrons form interference patterns when they are shot through slits in a screen.

Feynman's view of quantum reality is crucial in understanding the theories we will soon present, so it is worth taking some time to get a feeling for how it works. Imagine a simple process in which a particle begins at some location A and moves freely. In the Newtonian model that particle will follow a straight line. After a certain precise time passes, we will find the particle at some precise location B along that line. In Feynman's model a quantum particle samples every path connecting A and B, collecting a number called a phase for each path. That phase represents the position in the cycle of a wave, that is, whether the wave is at a crest or trough or some precise position in between. Feynman's mathematical prescription for calculating that phase showed that when you add together the waves from all the paths you get the "probability amplitude" that the particle, starting at A, will reach B. The square of that probability amplitude then gives the correct probability that the particle will reach B.

The phase that each individual path contributes to the Feynman sum (and hence to the probability of going from A to B) can be visualized as an arrow that is of fixed length but can point in any direction. To add two phases, you place the arrow representing one phase at the end of the arrow representing the other, to get a new arrow representing the sum. To add more phases, you simply continue the process. Note that when the phases line up, the arrow representing the total can be quite long. But if they point in different directions, they tend to cancel when you add them, leaving you with not much of an arrow at all. The idea is illustrated in the figures below.

To carry out Feynman's prescription for calculating the probability amplitude that a particle beginning at a location A will end up at a location B, you add the phases, or arrows, associated with every path connecting A and B. There are an infinite number of paths, which makes the mathematics a bit complicated, but it works. Some of the paths are pictured below.

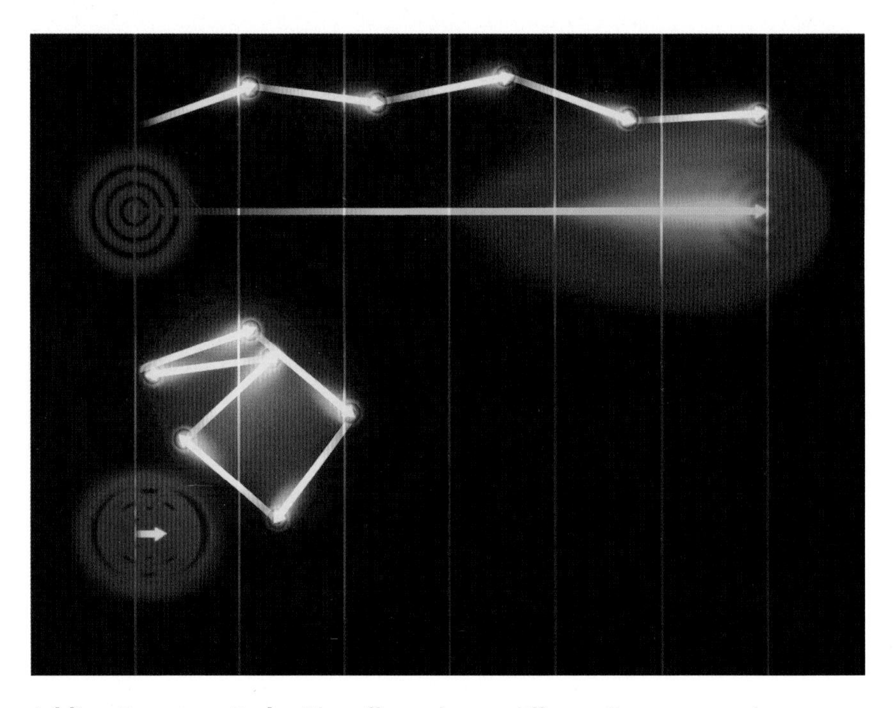

Adding Feynman Paths The effects due to different Feynman paths can enhance or diminish each other just as waves do. The yellow arrows represent the phases to be added. The blue lines represent their sum, a line from the tail of the first arrow to the point of the last one. In the lower image the arrows point in different directions and so their sum, the blue line, is very short.

Feynman's theory gives an especially clear picture of how a Newtonian world picture can arise from quantum physics, which seems very different. According to Feynman's theory, the phases associated with each path depend upon Planck's constant. The theory dictates that because Planck's constant is so small, when you add the contribution from paths that are close to each other

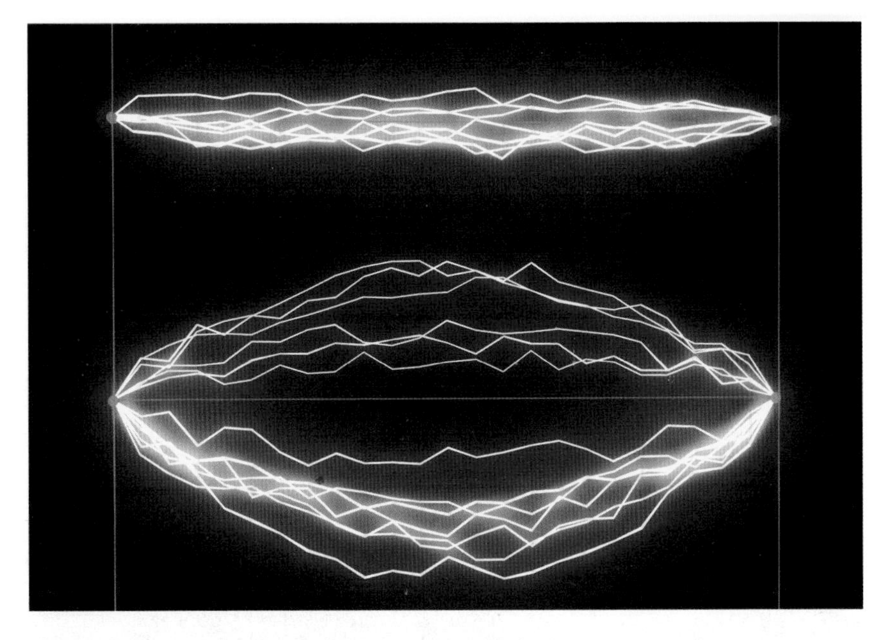

The Paths from A to B The "classical" path between two points is a straight line. The phases of paths that are near to the classical path tend to enhance each other, while the phases of paths farther from it tend to cancel out.

the phases normally vary wildly, and so, as in the figure above, they tend to add to zero. But the theory also shows that there are certain paths for which the phases have a tendency to line up, and so those paths are favoured; that is, they make a larger contribution to the observed behaviour of the particle. It turns out that for large objects, paths very similar to the path predicted by Newton's will have similar phases and add up to give by far the largest contribution to the sum, and so the only destination that has a probability effectively greater than zero is the destination predicted by Newtonian theory, and that destination has a probability that is very nearly one. Hence large objects move just as Newton's theory predicts they will.

So far we have discussed Feynman's ideas in the context of the double-slit experiment. In that experiment particles are fired toward a wall with slits, and we measure the location, on a screen placed beyond the wall, at which the particles end up. More generally, instead of just a single particle Feynman's theory allows us to predict the probable outcomes of a "system", which could be a particle, a set of particles, or even the entire universe. Between the initial state of the system and our later measurement of its properties, those properties evolve in some way, which physicists call the system's history. In the double-slit experiment, for example, the history of the particle is simply its path. Just as for the double-slit experiment the chance of observing the particle to land at any given point depends upon all the paths that could have gotten it there, Feynman showed that, for a general system, the probability of any observation is constructed from all the possible histories that could have led to that observation. Because of that his method is called the "sum over histories" or "alternative histories" formulation of quantum physics.

Now that we have a feeling for Feynman's approach to quantum physics, it is time to examine another key quantum principle that we will use later—the principle that observing a system must alter its course. Can't we, as we do when our supervisor has a spot of mustard on her chin, discreetly watch but not interfere? No. According to quantum physics, you cannot "just" observe something. That is, quantum physics recognizes that to make an observation, you must interact with the object you are observing. For instance, to see an object in the traditional sense, we shine a light on it. Shining a light on a pumpkin will of course have little effect on it. But shining even a dim light on a tiny quantum particle—that is, shooting photons at it—does have an appreciable effect, and experiments show that it changes the results of an experiment in just the way that quantum physics describes.

Suppose that, as before, we send a stream of particles towards the barrier in the double-slit experiment and collect data on the first million particles to get through. When we plot the number of particles landing at various detection points the data will form the interference pattern pictured on page 65, and when we add the phases associated with all the possible paths from a particle's starting point A to its detection point B, we will find that the probability we calculate of landing at various points agrees with that data.

Now suppose we repeat the experiment, this time shining lights on the slits so that we know an intermediate point, C, through which the particle passed. (C is the position of either one of the slits or the other.) This is called "which-path" information because it tells us whether each particle went from A to slit 1 to B, or from A to slit 2 to B. Since we now know through which slit each particle passed, the paths in our sum for that particle will now include only paths that travel through slit 1, or only paths that travel through slit 2. It will never include both the paths that go through slit 1 and the paths that pass through slit 2. Because Feynman explained the interference pattern by saying that paths that go through one slit interfere with paths that go through the other, if you turn on a light to determine which slit the particles pass through, thereby eliminating the other option, you will make the interference pattern disappear. And indeed, when the experiment is performed, turning on a light changes the results from the interference pattern on page 65, to a pattern like that on page 64! Moreover, we can vary the experiment by employing very faint light so that not all of the particles interact with the light. In that case we are able to obtain which-path information for only some subset of the particles. If we then divide the data on particle arrivals according to whether or not we obtained which-path information, we find that data pertaining to the subset for which we have no which-path information will form an interference pattern, and the subset of data pertaining to the particles for which we do have which-path information will not show interference.

This idea has important implications for our concept of "the past". In Newtonian theory, the past is assumed to exist as a definite series of events. If you see that vase you bought in Italy last year lying smashed on the floor and your toddler standing over it looking sheepish, you can trace backward the events that led to the mishap: the little fingers letting go, the vase falling and exploding into a thousand pieces as it hits. In fact, given complete data about the present, Newton's laws allow one to calculate a complete picture of the past. This is consistent with our intuitive understanding that, whether painful or joyful, the world has a definite past. There may have been no one watching, but the past exists as surely as if you had taken a series of snapshots of it. But a quantum buckyball cannot be said to have taken a definite path from source to screen. We might pin down a buckyball's location by observing it, but in between our observations, it takes all paths. Quantum physics tells us that no matter how thorough our observation of the present, the (unobserved) past, like the future, is indefinite and exists only as a spectrum of possibilities. The universe, according to quantum physics, has no single past, or history.

The fact that the past takes no definite form means that observations you make on a system in the present affect its past. That is underlined rather dramatically by a type of experiment thought up by physicist John Wheeler, called a delayed-choice experiment. Schematically, a delayed-choice experiment is like the double-slit experiment we just described, in which you have the option of observing the path that the particle takes, except in the delayed-choice experiment you postpone your decision about whether or not to observe the path until just before the particle hits the detection screen.

Delayed-choice experiments result in data identical to those we get when we choose to observe (or not observe) the which-path information by watching the slits themselves. But in this case the path each particle takes—that is, its past—is determined long after it passed through the slits and presumably had to "decide" whether to travel through just one slit, which does not produce interference, or both slits, which does.

Wheeler even considered a cosmic version of the experiment, in which the particles involved are photons emitted by powerful quasars billions of light-years away. Such light could be split into two paths and refocused toward earth by the gravitational lensing of an intervening galaxy. Though the experiment is beyond the reach of current technology, if we could collect enough photons from this light, they ought to form an interference pattern. Yet if we place a device to measure which-path information shortly before detection, that pattern should disappear. The choice whether to take one or both paths in this case would have been made billions of years ago, before the earth or perhaps even our sun was formed, and yet with our observation in the laboratory we will be affecting that choice.

In this chapter we have illustrated quantum physics employing the double-slit experiment. In what follows we will apply Feynman's formulation of quantum mechanics to the universe as a whole. We will see that, like a particle, the universe doesn't have just a single history, but every possible history, each with its own probability; and our observations of its current state affect its past and determine the different histories of the universe, just as the observations of the particles in the double-slit experiment affect the particles' past. That analysis will show how the laws of nature in our universe arose from the big bang. But before we examine how the laws arose, we'll talk a little bit about what those laws are, and some of the mysteries that they provoke.

5

THE THEORY OF EVERYTHING

The most incomprehensible thing about the universe is that it is comprehensible.

—Albert Einstein

THE UNIVERSE IS COMPREHENSIBLE because it is governed by scientific laws; that is to say, its behaviour can be modelled. But what are these laws or models? The first force to be described in mathematical language was gravity. Newton's law of gravity, published in 1687, said that every object in the universe attracts every other object with a force proportional to its mass. It made a great impression on the intellectual life of its era because it showed for the first time that at least one aspect of the universe could be accurately modelled, and it established the mathematical machinery to do so. The idea that there are laws of nature brings up issues similar to that for which Galileo had been convicted of heresy about fifty years earlier. For instance, the Bible tells the story of Joshua praying for the sun and moon to stop in their trajectories so he would have extra daylight to finish fighting the Amorites in Canaan. According to the book of Joshua, the sun stood still for about a day. Today we know that that would have meant that the earth stopped rotating. If the earth stopped, according to Newton's laws anything not tied down would have remained in motion at the earth's original speed (1,100 miles per hour at the equator)—a high price to pay for a delayed sunset. None of this bothered Newton himself, for as we've said, Newton believed that God could and did intervene in the workings of the universe.

The next aspects of the universe for which a law or model was discovered were the electric and magnetic forces. These behave like gravity, with the important difference that two electric charges or two magnets of the same kind repel each other, while unlike charges or unlike magnets attract. Electric and magnetic forces are far stronger than gravity, but we don't usually notice them in everyday life because a macroscopic body contains almost equal numbers of positive and negative electrical charges. This means that the electric and magnetic forces between two macroscopic bodies nearly cancel each other out, unlike the gravitational forces, which all add up.

Our current ideas about electricity and magnetism were developed over a period of about a hundred years from the mid-eighteenth to the mid-nineteenth century, when physicists in several countries made detailed experimental studies of electric and magnetic forces. One of the most important discoveries was that electrical and magnetic forces are related: a moving electrical charge causes a force on magnets, and a moving magnet causes a force on electrical charges. The first to realize there was some connection was Danish physicist Hans Christian Ørsted. While setting up for a lecture he was to give at the university in 1820, Ørsted noticed that the electric current from the battery he was using deflected a nearby compass needle. He soon realized that moving electricity created a magnetic force, and coined the term "electromagnetism". A few years later British scientist Michael Faraday reasoned that—expressed in modern terms—if an electric current could cause a magnetic field, a magnetic field should be able to produce an electric current. He demonstrated that effect in 1831. Fourteen years later Faraday also discovered a connection between electromagnetism and light when he showed that intense magnetism can affect the nature of polarized light.

Faraday had little formal education. He had been born into a poor blacksmith's family near London and left school at age thirteen to work as an errand boy and bookbinder in a bookshop. There, over the years, he learned science by reading the books he was supposed to care for, and by performing simple and cheap experiments in his spare time. Eventually he obtained work as an assistant in the laboratory of the great chemist Sir Humphry Davy. Faraday would stay on for the remaining forty-five years of his life and, after Davy's death, succeed him. Faraday had trouble with mathematics and never learned much of it, so it was a struggle for him to conceive a theoretical picture of the odd electromagnetic phenomena he observed in his laboratory. Nevertheless, he did.

One of Faraday's greatest intellectual innovations was the idea of force fields. These days, thanks to books and movies about bug-eyed aliens and their starships, most people are familiar with the term, so maybe he should get a royalty. But in the centuries between Newton and Faraday one of the great mysteries of physics was that its laws seemed to indicate that forces act across the empty space that separates interacting objects. Faraday didn't like that. He believed that to move an object, something has to come in contact with it. And so he imagined the space between electric charges and magnets as being filled with invisible tubes that physically do the pushing and pulling. Faraday called those tubes a force field. A good way to visualize a force field is to perform the schoolroom demonstration in which a glass plate is placed over a bar magnet and iron filings spread on the glass. With a few taps to overcome friction, the filings move as if nudged by an unseen power and arrange themselves in a pattern of arcs stretching from one pole of the magnet to the other. That pattern is a map of the unseen magnetic force that permeates space. Today we believe that all forces are transmitted by fields, so

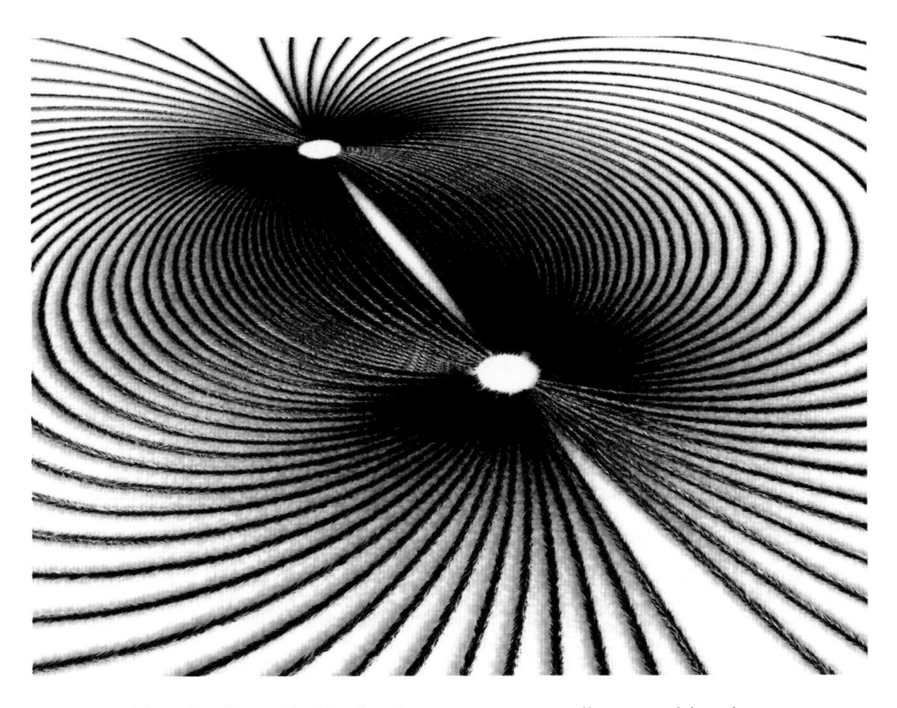

Force Fields The force field of a bar magnet, as illustrated by the reaction of iron filings.

it is an important concept in modern physics—as well as science fiction.

For several decades our understanding of electromagnetism remained stalled, amounting to no more than the knowledge of a few empirical laws: the hint that electricity and magnetism were closely, if mysteriously, related; the notion that they had some sort of connection to light; and the embryonic concept of fields. At least eleven theories of electromagnetism existed, every one of them flawed. Then, over a period of years in the 1860s, Scottish physicist James Clerk Maxwell developed Faraday's thinking into a mathematical framework that explained the intimate and mysterious relation among electricity, magnetism and light. The result was a set of equations describing both electric and magnetic forces as manifestations of the same physical entity, the electromagnetic field. Maxwell had unified electricity and magnetism into one force. Moreover, he showed that electromagnetic fields could propagate through space as a wave. The speed of that wave is governed by a number that appeared in his equations, which he calculated from experimental data that had been measured a few years earlier. To his astonishment the speed he calculated equalled the speed of light, which was then known experimentally to an accuracy of 1 percent. He had discovered that light itself is an electromagnetic wave!

Today the equations that describe electric and magnetic fields are called Maxwell's equations. Few people have heard of them, but they are probably the most commercially important equations we know. Not only do they govern the working of everything from household appliances to computers, but they also describe waves other than light, such as microwaves, radio waves, infrared light, and X-rays. All of these differ from visible light in only one respect—their wavelength. Radio waves have wavelengths of a metre or more, while visible light has a wavelength of a few ten-millionths of a metre, and X-rays a wavelength shorter than a hundred-millionth of a metre. Our sun radiates at all wavelengths, but its radiation is most intense in the wavelengths that are visible to us. It's probably no accident that the wavelengths we are able to see with the naked eye are those in which the sun radiates most strongly: it's likely that our eyes evolved with the ability to detect electromagnetic radiation in that range precisely because that is the range of radiation most available to them. If we ever run into beings from other planets, they will probably have the ability to "see" radiation at whatever wavelengths their own sun emits most strongly, modulated by factors such as the light-blocking characteristics of the dust and gases in their planet's atmosphere. So aliens who

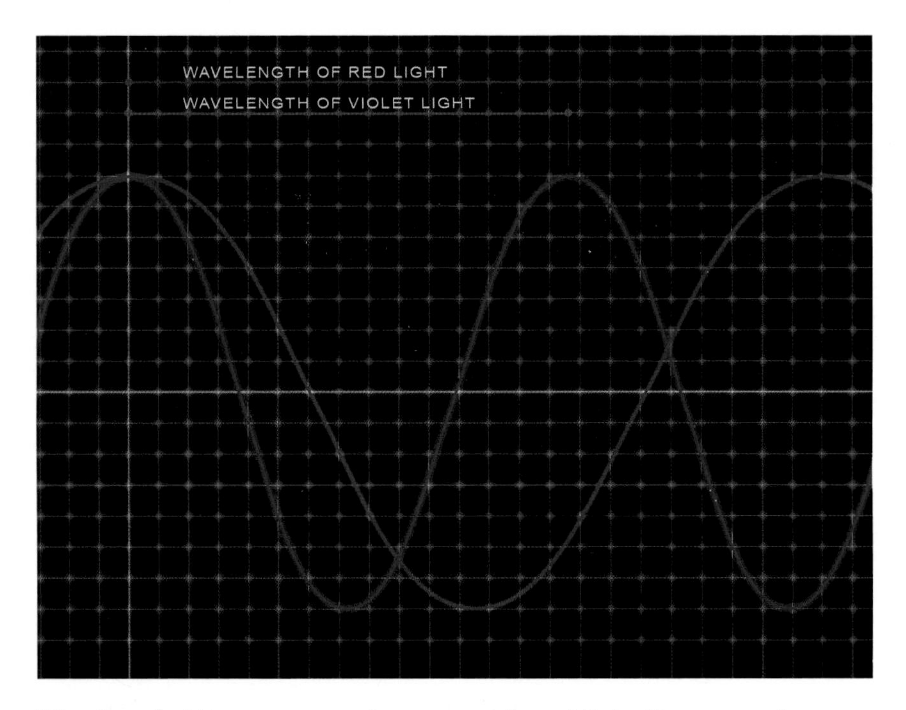

Wavelength Microwaves, radio waves, infrared light, X-rays—and different colours of light—differ only in their wavelengths.

evolved in the presence of X-rays might have a nice career in airport security.

Maxwell's equations dictate that electromagnetic waves travel at a speed of about 300,000 kilometres a second, or about 670 million miles per hour. But to quote a speed means nothing unless you specify a frame of reference relative to which the speed is measured. That's not something you usually need to think about in everyday life. When a speed limit sign reads 60 miles per hour, it is understood that your speed is measured relative to the road and not the black hole at the centre of the Milky Way. But even in everyday life there are occasions in which you have to take into account reference frames. For example, if you carry a cup of tea up the aisle of a jet plane in flight, you might say your speed is 2 miles per hour. Someone on the ground, however, might say you were moving at 572 miles per hour. Lest you think that one or the other of those observers has a better claim to the truth, keep in mind that because the earth orbits the sun, someone watching you from the surface of that heavenly body would disagree with both and say you are moving at about 18 miles per second, not to mention envying your air-conditioning. In light of such disagreements, when Maxwell claimed to have discovered the "speed of light" popping out of his equations, the natural question was, what is the speed of light in Maxwell's equations measured relative to?

There is no reason to believe that the speed parameter in Maxwell's equations is a speed measured relative to the earth. His equations, after all, apply to the entire universe. An alternative answer that was considered for a while is that his equations specify the speed of light relative to a previously undetected medium permeating all space, called the luminiferous ether, or for short, simply the ether, which was Aristotle's term for the substance he believed filled all of the universe outside the terrestrial sphere. This hypothetical ether would be the medium through which electromagnetic waves propagate, just as sound propagates through air. If the ether existed, there would be an absolute standard of rest (that is, rest with respect to the ether) and hence an absolute way of defining motion as well. The ether would provide a preferred frame of reference throughout the entire universe, against which any object's speed could be measured. So the ether was postulated to exist on theoretical grounds, setting some scientists off on a search for a way to study it, or at least to confirm its existence. One of those scientists was Maxwell himself.

If you race through the air toward a sound wave, the wave approaches you faster, and if you race away, it approaches you more slowly. Similarly, if there were an ether, the speed of light would vary depending on your motion relative to the ether. In fact, if light worked the way sound does, just as people on a supersonic jet will never hear any sound that emanates from behind the plane, so too would travellers racing quickly enough through the ether be able to outrun a light wave. Working from such considerations, Maxwell suggested an experiment. If there is an ether, the earth must be moving through it as it orbits the sun. And since the earth is travelling in a different direction in January than, say, in April or July, one ought to be able to observe a tiny difference in the speed of light at different times of the year—see the figure below.

Maxwell was talked out of publishing his idea in *Proceedings of*

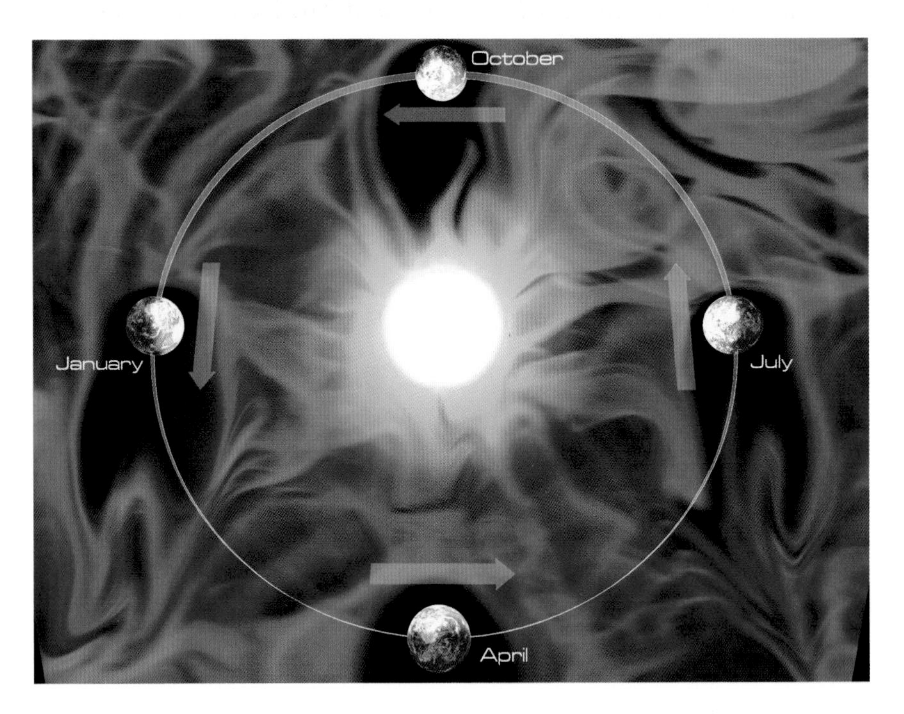

Moving Through the Ether If we were moving through the ether, we ought to be able to detect that motion by observing seasonal differences in the speed of light.

the Royal Society by its editor, who didn't think the experiment would work. But in 1879, shortly before he died at age forty-eight of painful stomach cancer, Maxwell sent a letter on the subject to a friend. The letter was published posthumously in the journal *Nature*, where it was read by, among others, an American physicist named Albert Michelson. Inspired by Maxwell's speculation, in 1887 Michelson and Edward Morley carried out a very sensitive experiment designed to measure the speed at which the earth travels through the ether. Their idea was to compare the speed of light in two different directions, at right angles. If the speed of light were a fixed number relative to the ether, the measurements should have revealed light speeds that differed depending on the direction of the beam. But Michelson and Morley observed no such difference.

The outcome of the Michelson and Morley experiment is clearly in conflict with the model of electromagnetic waves travelling through an ether, and should have caused the ether model to be abandoned. But Michelson's purpose had been to measure the speed of the earth relative to the ether, not to prove or disprove the ether hypothesis, and what he found did not lead him to conclude that the ether didn't exist. No one else drew that conclusion either. In fact, the famous physicist Sir William Thomson (Lord Kelvin) said in 1884 that the ether was "the only substance we are confident of in dynamics. One thing we are sure of, and that is the reality and substantiality of the luminiferous ether."

How can you believe in the ether despite the results of the Michelson-Morley experiment? As we've said often happens, people tried to save the model by contrived and ad hoc additions. Some postulated that the earth dragged the ether along with it, so we weren't actually moving with respect to it. Dutch physicist

Hendrik Antoon Lorentz and Irish physicist George Francis FitzGerald suggested that in a frame that was moving with respect to the ether, probably due to some yet-unknown mechanical effect, clocks would slow down and distances would shrink, so one would still measure light to have the same speed. Such efforts to save the aether concept continued for nearly twenty years until a remarkable paper by a young and unknown clerk in the patent office in Berne, Albert Einstein.

Einstein was twenty-six in 1905 when he published his paper "Zur Elektrodynamik bewegter Körper" ("On the Electrodynamics of Moving Bodies"). In it he made the simple assumption that the laws of physics and in particular the speed of light should appear to be the same to all uniformly moving observers. This idea, it turns out, demands a revolution in our concept of space and time. To see why, imagine two events that take place at the same spot but at different times, in a jet aircraft. To an observer on the jet there will be zero distance between those two events. But to a second observer on the ground the events will be separated by the distance the jet has travelled in the time between the events. This shows that two observers who are moving relative to each other will not agree on the distance between two events.

Now suppose the two observers observe a pulse of light travelling from the tail of the aircraft to its nose. Just as in the above example, they will not agree on the distance the light has travelled from its emission at the plane's tail to its reception at the nose. Since speed is distance travelled divided by the time taken, this means that if they agree on the speed at which the pulse travels—the speed of light—they will not agree on the time interval between the emission and the reception.

What makes this strange is that, though the two observers measure different times, they are watching the *same physical process*.

Airborne Jet If you bounce a ball on a jet, an observer aboard the plane may determine that it hits the same spot each bounce, while an observer on the ground will measure a large difference in the bounce points.

Einstein didn't attempt to construct an artificial explanation for this. He drew the logical, if startling, conclusion that the measurement of the time taken, like the measurement of the distance covered, depends on the observer doing the measuring. That effect is one of the keys to the theory in Einstein's 1905 paper, which has come to be called special relativity.

We can see how this analysis could apply to timekeeping devices if we consider two observers looking at a clock. Special relativity holds that the clock runs faster according to an observer who is at rest with respect to the clock. To observers who are not at rest with respect to the clock, the clock runs slower. If we liken a light pulse travelling from the tail to the nose of the plane to the

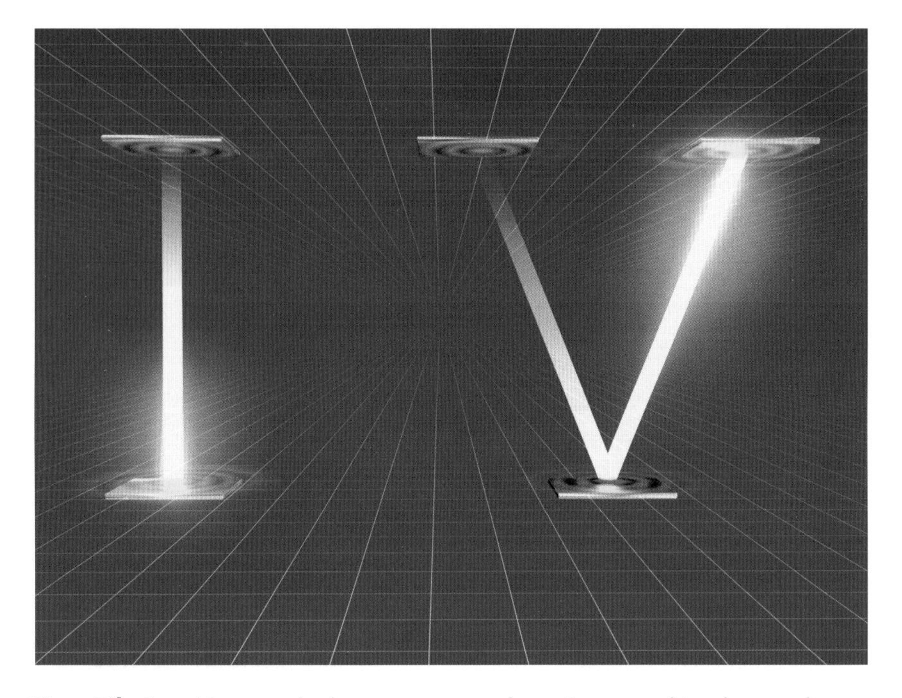

Time Dilation Moving clocks seem to run slow. Because this also applies to biological clocks, moving people will seem to age more slowly, but don't get your hopes up—at everyday speeds, no normal clock could measure the difference.

tick of a clock, we see that to an observer on the ground the clock runs slower because the light beam has to travel a greater distance in that frame of reference. But the effect does not depend on the mechanism of the clock; it holds for all clocks, even our own biological ones.

Einstein's work showed that, like the concept of rest, time cannot be absolute, as Newton thought. In other words, it is not possible to assign to every event a time with which every observer will agree. Instead, all observers have their own measures of time, and the times measured by two observers who are moving relative to each other will not agree. Einstein's ideas go counter to our intuition because their implications aren't noticeable at the speeds we normally encounter in everyday life. But they have been repeatedly confirmed by experiment. For example, imagine a reference clock at rest at the centre of the earth, another clock on the earth's surface, and a third clock aboard a plane, flying either with or against the direction of the earth's rotation. With reference to the clock at the earth's centre, the clock aboard the plane moving eastward—in the direction of the earth's rotation—is moving faster than the clock on the earth's surface, and so it should run slower. Similarly, with reference to the clock at the earth's centre, the clock aboard the plane flying westward—against the earth's rotation—is moving slower than the surface clock, which means that clock should run faster than the clock on the surface. And that is exactly what was observed when, in an experiment performed in October 1971, a very accurate atomic clock was flown around the world. So you could extend your life by constantly flying eastward around the world, though you might get tired of watching all those airline movies. However, the effect is very small, about 180 billionths of a second per circuit (and it is also somewhat lessened by the effects of the difference in gravity, but we need not get into that here).

Due to the work of Einstein, physicists realized that by demanding that the speed of light be the same in all frames of reference, Maxwell's theory of electricity and magnetism dictates that time cannot be treated as separate from the three dimensions of space. Instead, time and space are intertwined. It is something like adding a fourth direction of future/past to the usual left/right, forward/backward, and up/down. Physicists call this marriage of space and time "space-time", and because space-time includes a fourth direction, they call it the fourth dimension. In space-time, time is no longer separate from the three dimensions of space, and, loosely speaking, just as the definition of left/right, forward/backward, or up/down depends on the orientation of the observer, so too does the direction of time vary depending on the speed of the observer. Observers moving at different speeds would choose different directions for time in space-time. Einstein's theory of special relativity was therefore a new model, which got rid of the concepts of absolute time and absolute rest (i.e., rest with respect to the fixed ether).

Einstein soon realized that to make gravity compatible with relativity another change was necessary. According to Newton's theory of gravity, at any given time objects are attracted to each other by a force that depends on the distance between them at that time. But the theory of relativity had abolished the concept of absolute time, so there was no way to define when the distance between the masses should be measured. Thus Newton's theory of gravity was not consistent with special relativity and had to be modified. The conflict might sound like a mere technical difficulty, perhaps even a detail that could somehow be worked around without much change in the theory. As it turned out, nothing could have been further from the truth.

Over the next eleven years Einstein developed a new theory of gravity, which he called general relativity. The concept of gravity in general relativity is nothing like Newton's. Instead, it is based on the revolutionary proposal that space-time is not flat, as had been assumed previously, but is curved and distorted by the mass and energy in it.

A good way to picture curvature is to think of the surface of the earth. Although the earth's surface is only two-dimensional (because there are only two directions along it, say north/south and east/west), we're going to use it as our example because a curved two-dimensional space is easier to picture than a curved four-dimensional space. The geometry of curved spaces such as the earth's surface is not the Euclidean geometry we are familiar with. For example, on the earth's surface, the shortest distance between two points—which we know as a line in Euclidean geometry—is the path connecting the two points along what is called a great circle. (A great circle is a circle along the earth's surface whose centre coincides with the centre of the earth. The equator is an example of a great circle, and so is any circle obtained by rotating the equator along different diameters.)

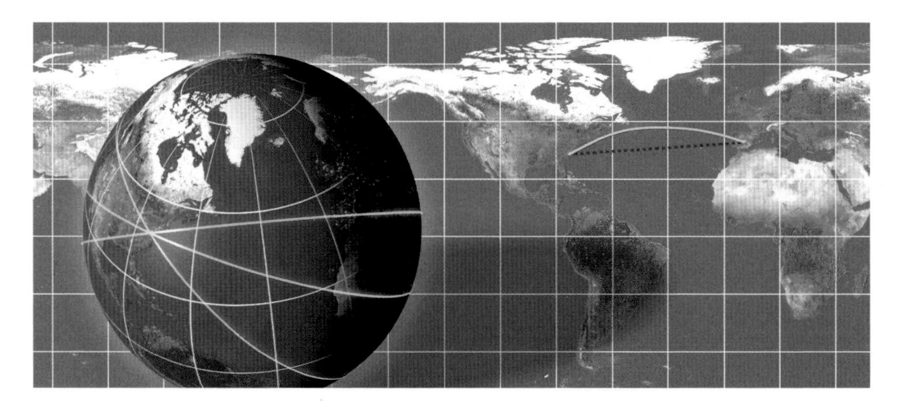

Geodesics The shortest distance between two points on the earth's surface appears curved when drawn on a flat map—something to keep in mind if ever given a sobriety test.

Imagine, say, that you wanted to travel from New York to Madrid, two cities that are at almost the same latitude. If the earth were flat, the shortest route would be to head straight east. If you did that, you would arrive in Madrid after travelling 3,707 miles. But due to the earth's curvature, there is a path that on a flat map looks curved and hence longer, but which is actually shorter. You can get there in 3,605 miles if you follow the great-circle route, which is to first head northeast, then gradually turn east, and then southeast. The difference in distance between the two routes is due to the earth's curvature, and a sign of its non-Euclidean geometry. Airlines know this, and arrange for their pilots to follow great-circle routes whenever practical.

According to Newton's laws of motion, objects such as cannonballs, croissants and planets move in straight lines unless acted upon by a force, such as gravity. But gravity, in Einstein's theory, is not a force like other forces; rather, it is a consequence of the fact that mass distorts space-time, creating curvature. In Einstein's theory, objects move on geodesics, which are the nearest things to straight lines in a curved space. Lines are geodesics on the flat plane, and great circles are geodesics on the surface of the earth. In the absence of matter, the geodesics in four-dimensional space-time correspond to lines in three-dimensional space. But when matter is present, distorting space-time, the paths of bodies in the corresponding three-dimensional space curve in a manner that in Newtonian theory was explained by the attraction of gravity. When space-time is not flat, objects' paths appear to be bent, giving the impression that a force is acting on them.

Einstein's general theory of relativity reproduces special relativity when gravity is absent, and it makes almost the same predictions as Newton's theory of gravity in the weak-gravity environment of our solar system—but not quite. In fact, if general relativity were not taken into account in GPS satellite navigation systems, errors in global positions would accumulate at a rate of about ten kilometres each day! However, the real importance of general relativity is not its application in devices that guide you to new restaurants, but rather that it is a very different model of the universe, which predicts new effects such as gravitational waves and black holes. And so general relativity has transformed physics into geometry. Modern technology is sensitive enough to allow us to perform many sensitive tests of general relativity, and it has passed every one.

Though they both revolutionized physics, Maxwell's theory of electromagnetism and Einstein's theory of gravity—general relativity—are both, like Newton's own physics, classical theories. That is, they are models in which the universe has a single history. As we saw in the last chapter, at the atomic and subatomic levels these models do not agree with observations. Instead, we have to use quantum theories in which the universe can have any possible history, each with its own intensity or probability amplitude. For practical calculations involving the everyday world, we can continue to use classical theories, but if we wish to understand the behaviour of atoms and molecules, we need a quantum version of Maxwell's theory of electromagnetism; and if we want to understand the early universe, when all the matter and energy in the universe were squeezed into a small volume, we must have a quantum version of the theory of general relativity. We also need such theories because if we are seeking a fundamental understanding of nature, it would not be consistent if some of the laws were quantum while others were classical. We therefore have to find quantum versions of all the laws of nature. Such theories are called quantum field theories.

The known forces of nature can be divided into four classes:

1. *Gravity*. This is the weakest of the four, but it is a long-range force and acts on everything in the universe as an attraction. This means that for large bodies the gravitational forces all add up and can dominate over all other forces.

2. *Electromagnetism*. This is also long-range and is much stronger than gravity, but it acts only on particles with an electric charge, being repulsive between charges of the same sign and attractive between charges of the opposite sign. This means the electric forces between large bodies cancel each other out, but on the

scales of atoms and molecules they dominate. Electromagnetic forces are responsible for all of chemistry and biology.

3. *Weak nuclear force.* This causes radioactivity and plays a vital role in the formation of the elements in stars and the early universe. We don't, however, come into contact with this force in our everyday lives.

4. *Strong nuclear force.* This force holds together the protons and neutrons inside the nucleus of an atom. It also holds together the protons and neutrons themselves, which is necessary because they are made of still tinier particles, the quarks we mentioned in Chapter 3. The strong force is the energy source for the sun and nuclear power, but, as with the weak force, we don't have direct contact with it.

The first force for which a quantum version was created was electromagnetism. The quantum theory of the electromagnetic field, called quantum electrodynamics, or QED for short, was developed in the 1940s by Richard Feynman and others, and has become a model for all quantum field theories. As we've said, according to classical theories, forces are transmitted by fields. But in quantum field theories the force fields are pictured as being made of various elementary particles called bosons, which are force-carrying particles that fly back and forth between matter particles, transmitting the forces. The matter particles are called fermions. Electrons and quarks are examples of fermions. The photon, or particle of light, is an example of a boson. It is the boson that transmits the electromagnetic force. What happens is that a matter particle, such as an electron, emits a boson, or force particle, and recoils from it, much as a cannon recoils after firing a cannonball. The force particle then collides with another matter particle and is absorbed, changing the motion of that particle. According to QED, all the interactions between charged particles—particles that feel the electromagnetic force—are described in terms of the exchange of photons.

The predictions of QED have been tested and found to match experimental results with great precision. But performing the mathematical calculations required by QED can be difficult. The problem, as we'll see below, is that when you add to the above framework of particle exchange the quantum requirement that one include all the histories by which an interaction can occur—for example, all the ways the force particles can be exchanged—the mathematics becomes complicated. Fortunately, along with inventing the notion of alternative histories—the way of thinking about quantum theories described in the last chapter—Feynman also developed a neat graphical method of accounting for the different histories, a method that is today applied not just to QED but to all quantum field theories.

Feynman's graphical method provides a way of visualizing each term in the sum over histories. Those pictures, called Feynman diagrams, are one of the most important tools of modern physics. In QED the sum over all possible histories can be represented as a sum over Feynman diagrams like those below, which represent some of the ways it is possible for two electrons to scatter off each other through the electromagnetic force. In these diagrams the solid lines represent the electrons and the wavy lines represent photons. Time is understood as progressing from bottom to top, and places where lines join correspond to photons being emitted or absorbed by an electron. Diagram (A) represents the two electrons approaching each other, exchanging a photon, and then continuing on their way. That is the simplest way in which two electrons can interact electromagnetically, but we must consider all possible histories. Hence we must also include diagrams like (B). That diagram also pictures two lines coming in—the approaching electrons—and two lines going out—the scattered ones—but in this diagram the electrons exchange two photons before flying off. The diagrams pictured are only a few of the possibilities; in fact, there are an infinite number of diagrams, which must be mathematically accounted for.

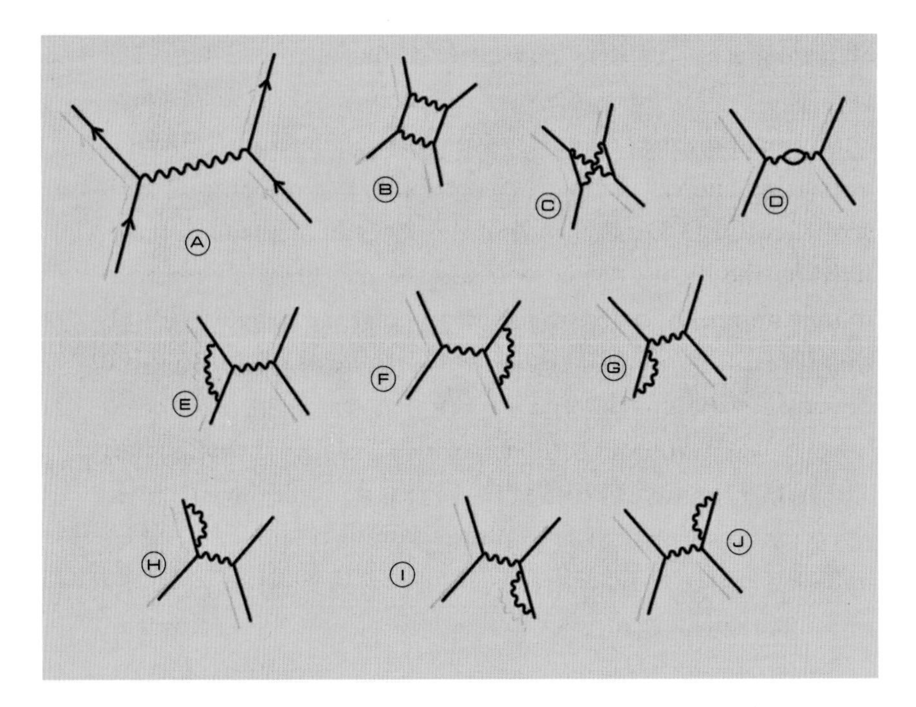

Feynman Diagrams These diagrams pertain to a process in which two electrons scatter off each other.

Feynman diagrams aren't just a neat way of picturing and categorizing how interactions can occur. Feynman diagrams come with rules that allow you to read off, from the lines and vertices in each diagram, a mathematical expression. The probability, say, that the incoming electrons, with some given initial momentum, will end up flying off with some particular final momentum is then obtained by summing the contributions from each Feynman diagram. That can take some work, because, as we've said, there are an infinite number of them. Moreover, although the incoming and outgoing electrons are assigned a definite energy and momentum, the particles in the closed loops in the interior of the diagram can have any energy and momentum. That is important because in forming the Feynman sum one must sum not only over all diagrams but also over all those values of energy and momentum.

Feynman diagrams provided physicists with enormous help in visualizing and calculating the probabilities of the processes described by QED. But they did not cure one important ailment suffered by the theory: When you add the contributions from the infinite number of different histories, you get an infinite result. (If the successive terms in an infinite sum decrease fast enough, it is possible for the sum to be finite, but that, unfortunately, doesn't happen here.) In particular, when the Feynman diagrams are added up, the answer seems to imply that the electron has an infinite mass and charge. This is absurd, because we can measure the mass and charge and they are finite. To deal with these infinities, a procedure called renormalization was developed.

The process of renormalization involves subtracting quantities that are defined to be infinite and negative in such a way that, with careful mathematical accounting, the sum of the negative infinite values and the positive infinite values that arise in the theory almost cancel out, leaving a small remainder, the finite observed values of mass and charge. These manipulations might sound like the sort of things that get you a flunking grade on a school maths exam, and renormalization is indeed, as it sounds, mathematically dubious. One consequence is that the values obtained by this

Feynman Diagrams Richard Feynman drove a famous van with Feynman diagrams painted on it. This artist's depiction was made to show the diagrams discussed above. Though Feynman died in 1988, the van is still around—in storage near Caltech in Southern California.

method for the mass and charge of the electron can be any finite number. That has the advantage that physicists may choose the negative infinities in a way that gives the right answer, but the disadvantage that the mass and charge of the electron therefore cannot be predicted from the theory. But once we have fixed the mass and charge of the electron in this manner, we can employ QED to make many other very precise predictions, which all agree extremely closely with observation, so renormalization is one of the essential ingredients of QED. An early triumph of QED, for example, was the correct prediction of the so-called Lamb shift, a small change in the energy of one of the states of the hydrogen atom discovered in 1947.

The success of renormalization in QED encouraged attempts to look for quantum field theories describing the other three forces of nature. But the division of natural forces into four classes is probably artificial and a consequence of our lack of understanding. People have therefore sought a theory of everything that will unify the four classes into a single law that is compatible with quantum theory. This would be the holy grail of physics.

One indication that unification is the right approach came from the theory of the weak force. The quantum field theory describing the weak force on its own cannot be renormalized; that is, it has infinities that cannot be cancelled by subtracting a finite number of quantities such as mass and charge. However, in 1967 Abdus Salam and Steven Weinberg each independently proposed a theory in which electromagnetism was unified with the weak force, and found that the unification cured the plague of infinities. The unified force is called the electroweak force. Its theory could be renormalized, and it predicted three new particles called W^+, W^-, and Z^0. Evidence for the Z^0 was discovered at CERN in Geneva in 1973. Salam and Weinberg were awarded the Nobel Prize in 1979, although the W and Z particles were not observed directly until 1983.

The strong force can be renormalized on its own in a theory called QCD, or quantum chromodynamics. According to QCD, the proton, the neutron, and many other elementary particles of matter are made of quarks, which have a remarkable property that physicists have come to call colour (hence the term "chromodynamics," although quark colours are just helpful labels—there is no connection with visible colour). Quarks come in three so-called colours, red, green and blue. In addition, each quark has an anti-particle partner, and the colours of those particles are called anti-red, anti-green and anti-blue. The idea is that only combinations with no net colour can exist as free particles. There are two ways to achieve such neutral quark combinations. A colour and its anti-colour cancel, so a quark and an anti-quark form a colourless pair, an unstable particle called a meson. Also, when all the three colours (or anti-colours) are mixed, the result has no net colour. Three quarks, one of each colour, form stable particles called baryons, of which protons and neutrons are examples (and three anti-quarks form the anti-particles of the baryons). Protons and neutrons are the baryons that make up the nucleus of atoms and are the basis for all normal matter in the universe.

QCD also has a property called asymptotic freedom, which we referred to, without naming it, in Chapter 3. Asymptotic freedom means that the strong forces between quarks are small when the quarks are close together but increase if they are farther apart, rather as though they were joined by rubber bands. Asymptotic freedom explains why we don't see isolated quarks in nature and have been unable to produce them in the laboratory. Still, even though we cannot observe individual quarks, we accept the model because it works so well at explaining the behaviour of protons, neutrons and other particles of matter.

After uniting the weak and electromagnetic forces, physicists in the 1970s looked for a way to bring the strong force into that theory. There are a number of so-called grand unified theories or GUTs that unify the strong forces with the weak force and electromagnetism, but they mostly predict that protons, the stuff that we are made of, should decay, on average, after about 10^{32} years. That is a very long lifetime, given that the universe is only about 10^{10} years old. But in quantum physics, when we say the average life-

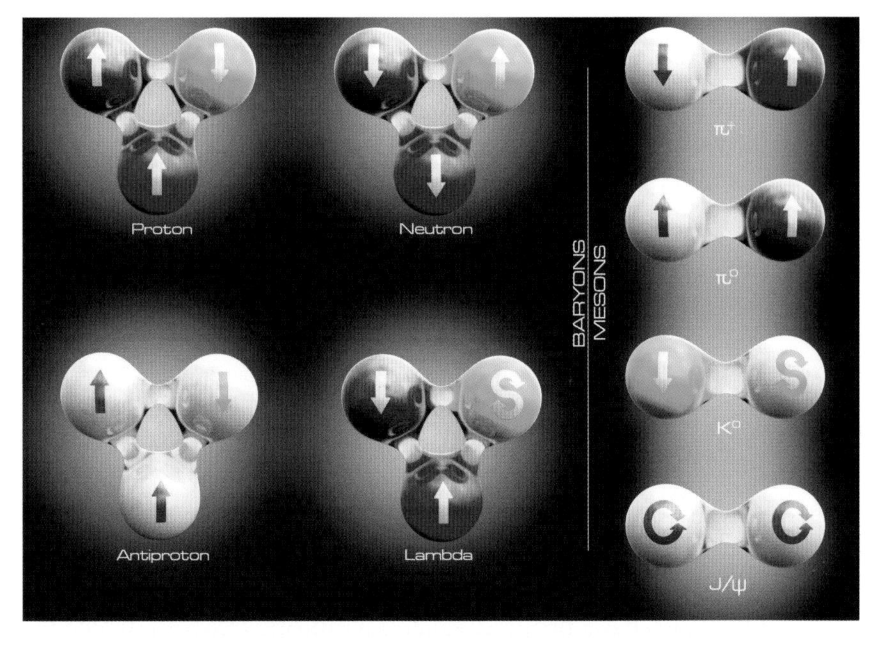

Baryons and Mesons Baryons and mesons are said to be made of quarks bound together by the strong force. When such particles collide, they can exchange quarks, but individual quarks cannot be observed.

time of a particle is 10^{32} years, we don't mean that most particles live approximately 10^{32} years, some a bit more and some a bit less. Instead, what we mean is that, each year, the particle has a 1 in 10^{32} chance of decaying. As a result, if you watch a tank containing 10^{32} protons for just a few years, you ought to see some of the protons decay. It is not too hard to build such a tank, since 10^{32} protons are contained in just a thousand tons of water. Scientists have performed such experiments. It turns out that detecting the decays and differentiating them from other events caused by the cosmic rays that continually shower us from space is no easy matter. To minimize the noise, the experiments are carried out deep inside places such as the Kamioka Mining and Smelting Company's mine 3,281 feet under a mountain in Japan, which is somewhat shielded from cosmic rays. As a result of observations in 2009, researchers have concluded that if protons decay at all, the proton lifetime is greater than about 10^{34} years, which is bad news for grand unified theories.

Since earlier observational evidence had also failed to support GUTs, most physicists adopted an ad hoc theory called the standard model, which comprises the unified theory of the electroweak forces and QCD as a theory of the strong forces. But in the standard model, the electroweak and strong forces act separately and are not truly unified. The standard model is very successful and agrees with all current observational evidence, but it is ultimately unsatisfactory because, apart from not unifying the electroweak and strong forces, it does not include gravity.

It may have proved difficult to meld the strong force with the electromagnetic and weak forces, but those problems are nothing compared with the problem of merging gravity with the other three, or even of creating a stand-alone quantum theory of gravity.

"Putting a box around it, I'm afraid, does *not* make it a unified theory."

The reason a quantum theory of gravity has proven so hard to create has to do with the Heisenberg uncertainty principle, which we discussed in Chapter 4. It is not obvious, but it turns out that with regard to that principle, the value of a field and its rate of change play the same role as the position and velocity of a particle. That is, the more accurately one is determined, the less accurately the other can be. An important consequence of that is that there is no such thing as empty space. That is because empty space means that both the value of a field and its rate of change are exactly zero. (If the field's rate of change were not zero, the space would not remain empty.) Since the uncertainty principle does not allow the values of both the field and the rate of change to be exact, space is never empty. It can have a state of minimum energy, called the vacuum, but that state is subject to what are called quantum jitters, or vacuum fluctuations—particles and fields quivering in and out of existence.

One can think of the vacuum fluctuations as pairs of particles that appear together at some time, move apart, then come together and annihilate each other. In terms of Feynman diagrams, they correspond to closed loops. These particles are called virtual particles. Unlike real particles, virtual particles cannot be observed directly with a particle detector. However, their indirect effects, such as small changes in the energy of electron orbits, can be measured, and agree with theoretical predictions to a remarkable degree of accuracy. The problem is that the virtual particles have energy, and because there are an infinite number of virtual pairs, they would have an infinite amount of energy. According to general relativity, this means that they would curve the universe to an infinitely small size, which obviously does not happen!

This plague of infinities is similar to the problem that occurs in the theories of the strong, weak, and electromagnetic forces, except in those cases renormalization removes the infinities. But the closed loops in the Feynman diagrams for gravity produce infinities that cannot be absorbed by renormalization because in general relativity there are not enough renormalizable parameters (such as the values of mass and charge) to remove all the quantum infinities from the theory. We are therefore left with a theory of gravity that predicts that certain quantities, such as the curvature of space-time, are infinite, which is no way to run a habitable universe. That means the only possibility of obtaining a sensible theory would be for all the infinities to somehow cancel, without resorting to renormalization.

In 1976 a possible solution to that problem was found. It is called supergravity. The prefix "super" was not appended because physicists thought it was "super" that this theory of quantum gravity might actually work. Instead, "super" refers to a kind of symmetry the theory possesses, called supersymmetry.

In physics a system is said to have a symmetry if its properties are unaffected by a certain transformation such as rotating it in space or taking its mirror image. For example, if you flip a doughnut over, it looks exactly the same (unless it has a chocolate topping, in which case it is better just to eat it). Supersymmetry is a more subtle kind of symmetry that cannot be associated with a transformation of ordinary space. One of the important implications of supersymmetry is that force particles and matter particles, and hence force and matter, are really just two facets of the same thing. Practically speaking, that means that each matter particle, such as a quark, ought to have a partner particle that is a force particle, and each force particle, such as the photon, ought to have a partner particle that is a matter particle. This has the potential to solve the problem of infinities because it turns out that the infinities from closed loops of force particles are positive while the infinities from closed loops of matter particles are negative, so the infinities in the theory arising from the force particles and their partner matter particles tend to cancel out. Unfortunately, the calculations required to find out whether there would be any infinities left uncancelled in supergravity were so long and difficult and had such potential for error that no one was prepared to undertake them. Most physicists believed, nonetheless, that supergravity was probably the right answer to the problem of unifying gravity with the other forces.

You might think that the validity of supersymmetry would be an easy thing to check—just examine the properties of the existing particles and see if they pair up. No such partner particles have been observed. But various calculations that physicists have performed indicate that the partner particles corresponding to the particles we observe ought to be a thousand times as massive as a proton, if not even heavier. That is too heavy for such particles to have been seen in any experiments to date, but there is hope that such particles will eventually be created in the Large Hadron Collider in Geneva.

The idea of supersymmetry was the key to the creation of supergravity, but the concept had actually originated years earlier with theorists studying a fledgling theory called string theory. According to string theory, particles are not points, but patterns of vibration that have length but no height or width—like infinitely thin pieces of string. String theories also lead to infinities, but it is believed that in the right version they will all cancel out. They have another unusual feature: they are consistent only if spacetime has ten dimensions, instead of the usual four. Ten dimensions might sound exciting, but they would cause real problems if you forgot where you parked your car. If they are present, why don't we notice these extra dimensions? According to string theory, they are curved up into a space of very small size. To picture this, imagine a two-dimensional plane. We call the plane two-dimensional because you need two numbers (for instance, horizontal and vertical coordinates) to locate any point on it. Another two-dimensional space is the surface of a straw. To locate a point on that space, you need to know where along the straw's length the point is, and also where along its circular dimension. But if the straw is very thin, you would get a very good approximate position employing only the coordinate that runs along the straw's length, so you might ignore the circular dimension. And if the straw were a million-million-million-million-millionth of an inch in diameter, you wouldn't notice the circular dimension at all. That is the picture string theorists have of the extra dimensions—they are highly curved, or curled, on a scale so small that we don't see them. In string theory the extra dimensions are curled up into what is called the internal space, as opposed to the three-dimensional space that we experience in everyday life. As we'll see, these internal states are not just hidden dimensions swept under the rug—they have important physical significance.

In addition to the question of dimensions, string theory suffered from another awkward issue: there appeared to be at least five different theories and millions of ways the extra dimensions could be curled up, which was quite an embarrassment of possibilities for those advocating that string theory was the *unique* theory of everything. Then, around 1994, people started to discover dualities—that different string theories, and different ways of curling up the extra dimensions, are simply different ways of describing the same phenomena in four dimensions. Moreover, they found that supergravity is also related to the other theories in this way. String theorists are now convinced that the five different string theories and supergravity are just different approximations to a more fundamental theory, each valid in different situations.

That more fundamental theory is called M-theory, as we mentioned earlier. No one seems to know what the "M" stands for, but it may be "master", "miracle" or "mystery". It seems to be all three. People are still trying to decipher the nature of M-theory, but that may not be possible. It could be that the physicist's traditional expectation of a single theory of nature is untenable, and there exists no single formulation. It might be that to describe the universe, we have to employ different theories in different situations. Each theory may have its own version of reality, but according to model-dependent realism, that is acceptable so long as the theories agree in their predictions whenever they overlap, that is, whenever they can both be applied.

Whether M-theory exists as a single formulation or only as a network, we do know some of its properties. First, M-theory has eleven space-time dimensions, not ten. String theorists had long suspected that the prediction of ten dimensions might have to be adjusted, and recent work showed that one dimension had indeed

Straws and Lines A straw is two-dimensional, but if its diameter is small enough—or if it is viewed from a distance—it appears to be one-dimensional, like a line.

been overlooked. Also, M-theory can contain not just vibrating strings but also point particles, two-dimensional membranes, three-dimensional blobs, and other objects that are more difficult to picture and occupy even more dimensions of space, up to nine. These objects are called p-branes (where p runs from zero to nine).

What about the enormous number of ways to curl up the tiny dimensions? In M-theory those extra space dimensions cannot be curled up in just any way. The mathematics of the theory restricts the manner in which the dimensions of the internal space can be curled. The exact shape of the internal space determines both the values of physical constants, such as the charge of the electron, and the nature of the interactions between elementary particles. In other words, it determines the apparent laws of nature. We say "apparent" because we mean the laws that we observe in our universe—the laws of the four forces, and the parameters such as mass and charge that characterize the elementary particles. But the more fundamental laws are those of M-theory.

The laws of M-theory therefore allow for *different universes* with different apparent laws, depending on how the internal space is curled. M-theory has solutions that allow for many different internal spaces, perhaps as many as 10^{500}, which means it allows for 10^{500} different universes, each with its own laws. To get an idea how many that is, think about this: If some being could analyse the laws predicted for each of those universes in just one millisecond and had started working on it at the big bang, at present that being would have studied just 10^{20} of them. And that's without coffee breaks.

Centuries ago Newton showed that mathematical equations could provide a startlingly accurate description of the way objects interact, both on earth and in the heavens. Scientists were led to believe that the future of the entire universe could be laid out if only we knew the proper theory and had enough computing power. Then came quantum uncertainty, curved space, quarks, strings and extra dimensions, and the net result of their labour is 10^{500} universes, each with different laws, only one of which corresponds to the universe as we know it. The original hope of physicists to produce a single theory explaining the apparent laws of our universe as the unique possible consequence of a few simple assumptions may have to be abandoned. Where does that leave us? If M-theory allows for 10^{500} sets of apparent laws, how did we end up in this universe, with the laws that are apparent to us? And what about those other possible worlds?

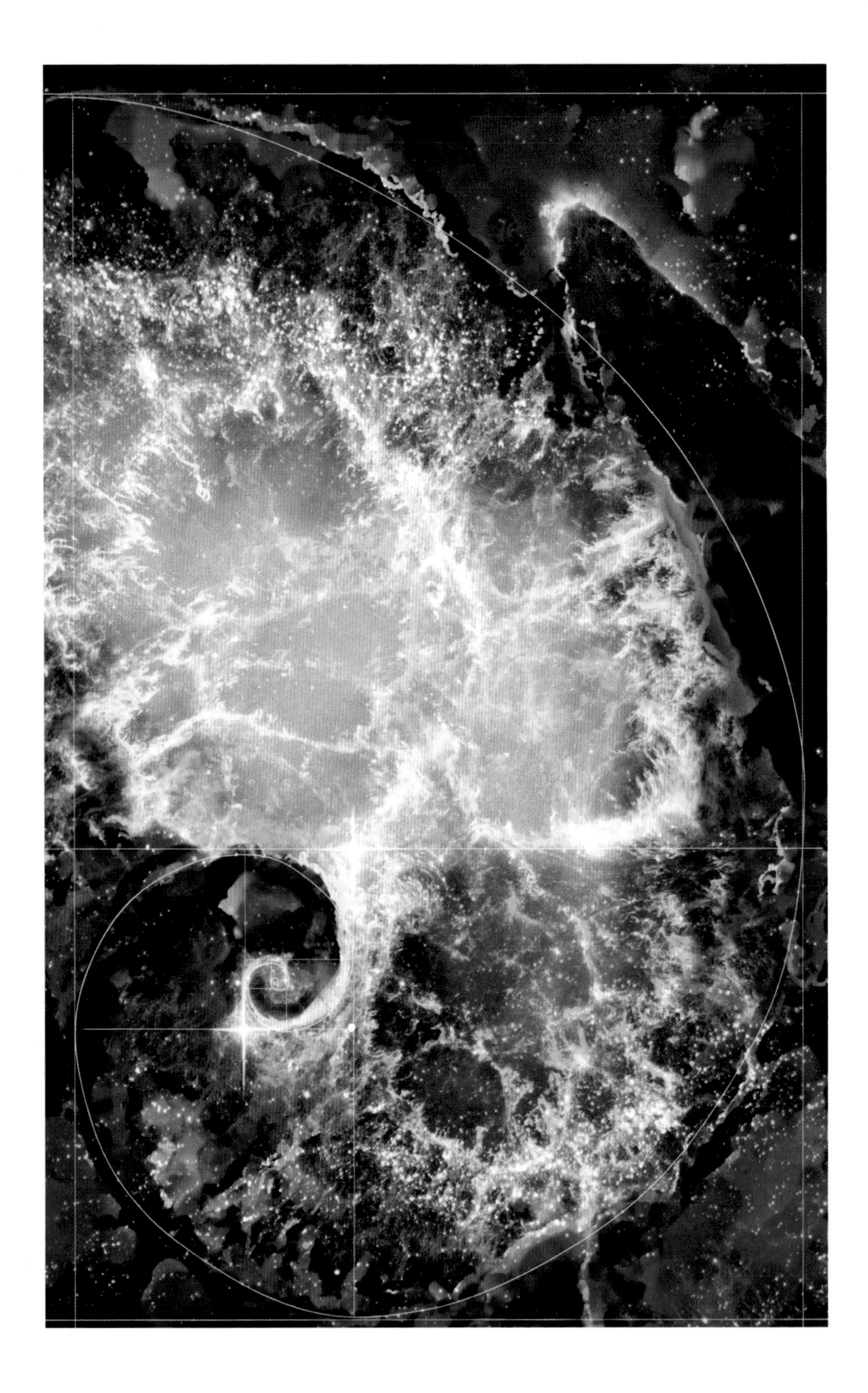

6

CHOOSING OUR UNIVERSE

According to the Boshongo people of central Africa, in the beginning there was only darkness, water and the great god Bumba. One day Bumba, in pain from a stomachache, vomited up the sun. In time the sun dried up some of the water, leaving land. But Bumba was still in pain, and vomited some more. Up came the moon, the stars, and then some animals: the leopard, the crocodile, the turtle and finally man. The Mayans of Mexico and Central America tell of a similar time before creation when all that existed were the sea, the sky and the Maker. In the Mayan legend the Maker, unhappy because there was no one to praise him, created the earth, mountains, trees and most animals. But the animals could not speak, and so he decided to create humans. First he made them of mud and earth, but they only spoke nonsense. He let them dissolve away and tried again, this time fashioning people from wood. Those people were dull. He decided to destroy them, but they escaped into the forest, sustaining damage along the way that altered them slightly, creating what we today know as monkeys. After that fiasco, the Maker finally came upon a formula that worked, and constructed the first humans from white and yellow corn. Today we make ethanol from corn, but so far haven't matched the Maker's feat of constructing the people who drink it.

Creation myths like these all attempt to answer the questions we address in this book: why is there a universe, and why is the universe the way it is? Our ability to address such questions has grown steadily in the centuries since the ancient Greeks, most profoundly over the past century. Armed with the background of the previous chapters, we are now ready to offer a possible answer to these questions.

One thing that may have been apparent even in early times was that either the universe was a very recent creation or else human beings have existed for only a small fraction of cosmic history. That's because the human race has been improving so rapidly in knowledge and technology that if people had been around for millions of years, the human race would be much further along in its mastery.

According to the Old Testament, God created Adam and Eve only six days into creation. Bishop Ussher, primate of all Ireland from 1625 to 1656, placed the origin of the world even more precisely, at nine in the morning on October 27, 4004 BC. We take a different view: that humans are a recent creation but that the universe itself began much earlier, about 13.7 billion years ago.

The first actual scientific evidence that the universe had a beginning came in the 1920s. As we said in Chapter 3, that was a time when most scientists believed in a static universe that had always existed. The evidence to the contrary was indirect, based upon the observations Edwin Hubble made with the 100-inch telescope on Mount Wilson, in the hills above Pasadena, California. By analysing the spectrum of light they emit, Hubble determined that nearly all galaxies are moving away from us, and the farther away they are, the faster they are moving. In 1929 he published a law relating their rate of recession to their distance from us, and concluded that the universe is expanding. If that is true, then the universe must have been smaller in the past. In fact, if we extrapolate to the distant past, all the matter and energy in the universe would have been concentrated in a very tiny region of unimaginable density and temperature, and if we go back far enough, there would be a time when it all began—the event we now call the big bang.

The idea that the universe is expanding involves a bit of subtlety. For example, we don't mean the universe is expanding in the manner that, say, one might expand one's house, by knocking out a wall and positioning a new bathroom where once there stood a majestic oak. Rather than space *extending* itself, it is the distance between any two points *within* the universe that is growing. That idea emerged in the 1930s amid much controversy, but one of the best ways to visualize it is still a metaphor enunciated in 1931 by Cambridge University astronomer Arthur Eddington. Eddington visualized the universe as the surface of an expanding balloon, and all the galaxies as points on that surface. This picture clearly illustrates why far galaxies recede more quickly than nearby ones. For example, if the radius of the balloon doubled each hour, then the distance between any two galaxies on the balloon would double each hour. If at some time two galaxies were 1 inch apart, an hour later they would be 2 inches apart, and they would appear to be moving relative to each other at a rate of 1 inch per hour. But if they started 2 inches apart, an hour later they would be separated by 4 inches and would appear to be moving away from each other at a rate of 2 inches per hour. That is just what Hubble found: the farther away a galaxy, the faster it was moving away from us.

It is important to realize that the expansion of space does not affect the size of material objects such as galaxies, stars, apples, atoms or other objects held together by some sort of force. For example, if we circled a cluster of galaxies on the balloon, that circle would not expand as the balloon expanded. Rather, because the galaxies are bound by gravitational forces, the circle and the galaxies within it would keep their size and configuration as the balloon enlarged. This is important because we can detect expansion only if our measuring instruments have fixed sizes. If everything were free to expand, then we, our yardsticks, our laboratories, and so on

Balloon Universe Distant galaxies recede from us as if the cosmos were all on the surface of a giant balloon.

would all expand proportionately and we would not notice any difference.

That the universe is expanding was news to Einstein. But the possibility that the galaxies are moving away from each other had been proposed a few years before Hubble's papers on theoretical grounds arising from Einstein's own equations. In 1922, Russian physicist and mathematician Alexander Friedmann investigated what would happen in a model universe based upon two assumptions that greatly simplified the mathematics: that the universe looks identical in every direction, and that it looks that way from every observation point. We know that Friedmann's first assumption is not exactly true—the universe fortunately is not uniform everywhere! If we gaze upward in one direction, we might see the sun; in another, the moon or a colony of migrating vampire bats.

But the universe does appear to be roughly the same in every direction when viewed on a scale that is far larger—larger even than the distance between galaxies. It is something like looking down at a forest. If you are close enough, you can make out individual leaves, or at least trees, and the spaces between them. But if you are so high up that if you hold out your thumb it covers a square mile of trees, the forest will appear to be a uniform shade of green. We would say that, on that scale, the forest is uniform.

Based on his assumptions Friedmann was able to discover a solution to Einstein's equations in which the universe expanded in the manner that Hubble would soon discover to be true. In particular, Friedmann's model universe begins with zero size and expands until gravitational attraction slows it down, and eventually causes it to collapse in upon itself. (There are, it turns out, two other types of solutions to Einstein's equations that also satisfy the assumptions of Friedmann's model, one corresponding to a universe in which the expansion continues forever, though it does slow a bit, and another to a universe in which the rate of expansion slows toward zero, but never quite reaches it.) Friedmann died a few years after producing this work, and his ideas remained largely unknown until after Hubble's discovery. But in 1927 a professor of physics and Roman Catholic priest named Georges Lemaître proposed a similar idea: if you trace the history of the universe backward into the past, it gets tinier and tinier until you come upon a creation event—what we now call the big bang.

Not everyone liked the big bang picture. In fact, the term "big bang" was coined in 1949 by Cambridge astrophysicist Fred Hoyle, who believed in a universe that expanded forever, and meant the term as a derisive description. The first direct observations supporting the idea didn't come until 1965, with the discovery that there is a faint background of microwaves throughout space. This cosmic microwave background radiation, or CMBR, is the same as that in your microwave oven, but much less powerful. You can observe the CMBR yourself by tuning your television to an unused channel—a few percent of the snow you see on the screen will be caused by it. The radiation was discovered by accident by two Bell Labs scientists trying to eliminate such static from their microwave antenna. At first they thought the static might be coming from the droppings of pigeons roosting in their apparatus, but it turned out their problem had a more interesting origin—the CMBR is radiation left over from the very hot and dense early universe that would have existed shortly after the big bang. As the universe expanded, it cooled until the radiation became just the faint remnant we now observe. At present these microwaves could heat your food to only about –270 degrees Centigrade—3 degrees above absolute zero, and not very useful for popping corn.

Astronomers have also found other fingerprints supporting the big bang picture of a hot, tiny early universe. For example, during the first minute or so, the universe would have been hotter than the centre of a typical star. During that period the entire universe would have acted as a nuclear fusion reactor. The reactions would have ceased when the universe expanded and cooled sufficiently, but the theory predicts that this should have left a universe composed mainly of hydrogen, but also about 23 percent helium, with traces of lithium (all heavier elements were made later, inside stars). The calculation is in good accordance with the amounts of helium, hydrogen and lithium we observe.

Measurements of helium abundance and the CMBR provided convincing evidence in favour of the big bang picture of the very early universe, but although one can think of the big bang picture as a valid description of early times, it is wrong to take the big bang literally, that is, to think of Einstein's theory as providing a true picture of the *origin* of the universe. That is because general relativity predicts there to be a point in time at which the temperature, density and curvature of the universe are all infinite, a situation mathematicians call a singularity. To a physicist this means that Einstein's theory breaks down at that point and therefore cannot be used to predict how the universe began, only how it evolved afterwards. So although we can employ the equations of general relativity and our observations of the heavens to learn about the universe at a very young age, it is not correct to carry the big bang picture all the way back to the beginning.

We will get to the issue of the origin of the universe shortly, but first a few words about the first phase of the expansion. Physicists call it inflation. Unless you've lived in Zimbabwe, where currency inflation recently exceeded 200,000,000 percent, the term may not sound very explosive. But according to even conservative estimates, during this cosmological inflation, the universe expanded by a factor of 1,000,000,000,000,000,000,000,000,000,000 in .00000000000000000000000000000000001 second. It was as if a coin 1 centimetre in diameter suddenly blew up to ten million times the width of the Milky Way. That may seem to violate relativity, which dictates that nothing can move faster than light, but that speed limit does not apply to the expansion of space itself.

The idea that such an episode of inflation might have occurred was first proposed in 1980, based on considerations that go beyond Einstein's theory of general relativity and take into account aspects of quantum theory. Since we don't have a complete quantum theory of gravity, the details are still being worked out, and physicists aren't sure exactly how inflation happened. But according to the theory, the expansion caused by inflation would not be *completely* uniform, as predicted by the traditional big bang picture. These irregularities would produce minuscule variations in the temperature of the CMBR in different directions. The variations are too small to have been observed in the 1960s, but they were first discovered in 1992 by NASA's COBE satellite, and later measured by its successor, the WMAP satellite, launched in 2001. As a result, we are now confident that inflation really did happen.

Ironically, though tiny variations in the CMBR are evidence for inflation, one reason inflation is an important concept is the nearly perfect uniformity of the temperature of the CMBR. If you make one part of an object hotter than its surroundings and then wait, the hot spot will grow cooler and its surroundings warmer until the temperature of the object is uniform. Similarly, one would expect the universe to eventually have a uniform temperature. But this process takes time, and if inflation hadn't occurred, there wouldn't have been enough time in the history of the universe for heat at widely separated regions to equalize, assuming that the speed of such heat transfer is limited by the speed of light. A period of very rapid expansion (much faster than light speed) remedies that because there would have been enough time for the equalization to happen in the extremely tiny preinflationary early universe.

Inflation explains the bang in the big bang, at least in the sense that the expansion it represents was far more extreme than the expansion predicted by the traditional big bang theory of general relativity during the time interval in which inflation occurred. The problem is, for our theoretical models of inflation to work, the initial state of the universe had to be set up in a very special and highly improbable way. Thus traditional inflation theory resolves one set of issues but creates another—the need for a very special initial state. That time-zero issue is eliminated in the theory of the creation of the universe we are about to describe.

Since we cannot describe creation employing Einstein's theory of general relativity, if we want to describe the origin of the universe, general relativity has to be replaced by a more complete theory. One would expect to need a more complete theory even if general relativity did not break down, because general relativity does not take into account the small-scale structure of matter, which is governed by quantum theory. We mentioned in Chapter 4 that for most practical purposes quantum theory does not hold much relevance for the study of the large-scale structure of the universe because quantum theory applies to the description of nature on microscopic scales. But if you go far enough back in time, the universe was as small as the Planck size, a billion-trillion-trillionth of a centimetre, which is the scale at which quantum theory does have to be taken into account. So though we don't yet have a complete quantum theory of gravity, we do know that the origin of the universe was a quantum event. As a result, just as we combined quantum theory and general relativity—at least provisionally—to derive the theory of inflation, if we want to go back even further and understand the origin of the universe, we must combine what we know about general relativity with quantum theory.

To see how this works, we need to understand the principle that gravity warps space and time. Warpage of space is easier to visualize than warpage of time. Imagine that the universe is the surface of a flat billiard table. The table's surface is a flat space, at least in two dimensions. If you roll a ball on the table it will travel in a straight line. But if the table becomes warped or dented in places, as in the illustration below, then the ball will curve.

It is easy to see how the billiard table is warped in this example

Space Warp Matter and energy warp space, altering the paths of objects.

because it is curving into an outside third dimension, which we can see. Since we can't step outside our own space-time to view its warpage, the space-time warpage in our universe is harder to imagine. But curvature can be detected even if you cannot step out and view it from the perspective of a larger space. It can be detected from within the space itself. Imagine a micro-ant confined to the surface of the table. Even without the ability to leave the table, the ant could detect the warpage by carefully charting distances. For example, the distance around a circle in flat space is always a bit more than three times the distance across its diameter (the actual multiple is π). But if the ant cut across a circle encompassing the well in the table pictured above, it would find that the distance across is greater than expected, greater than one-third the distance around it. In fact, if the well were deep enough, the ant would find that the distance around the circle is *shorter* than the distance across it. The same is true of warpage in our universe—it stretches or compresses the distance between points of space, changing its geometry, or shape, in a way that is measurable from within the universe. Warpage of time stretches or compresses time intervals in an analogous manner.

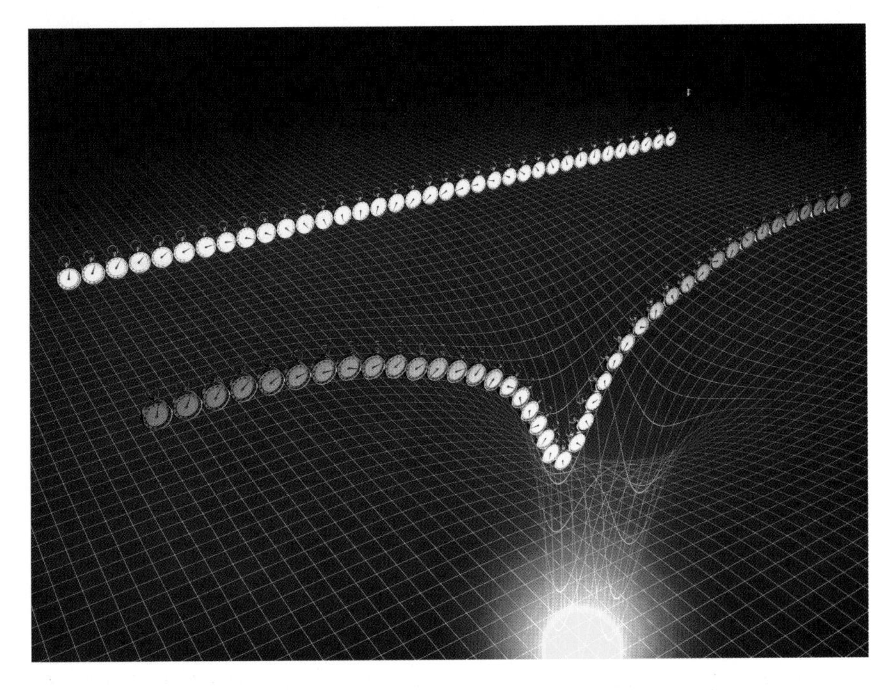

Space-time Warp Matter and energy warp time and cause the time dimension to "mix" with the space dimensions.

Armed with these ideas, let's return to the issue of the beginning of the universe. We can speak separately of space and time, as we have in this discussion, in situations involving low speeds and weak gravity. In general, however, time and space can become intertwined, and so their stretching and compressing also involve a certain amount of mixing. This mixing is important in the early universe and the key to understanding the beginning of time.

The issue of the beginning of time is a bit like the issue of the edge of the world. When people thought the world was flat, one might have wondered whether the sea poured over its edge. This has been tested experimentally: one can go around the world and not fall off. The problem of what happens at the edge of the world was solved when people realized that the world was not a flat plate, but a curved surface. Time, however, seemed to be like a model railway track. If it had a beginning, there would have to have been someone (i.e. God) to set the trains going. Although Einstein's general theory of relativity unified time and space as space-time and involved a certain mixing of space and time, time was still different from space, and either had a beginning and an end or else went on forever. However, once we add the effects of quantum theory to the theory of relativity, in extreme cases warpage can occur to such a great extent that time behaves like another dimension of space.

In the early universe—when the universe was small enough to be governed by both general relativity and quantum theory—there were effectively four dimensions of space and none of time. That means that when we speak of the "beginning" of the universe, we are skirting the subtle issue that as we look backwards towards the very early universe, time as we know it does not exist! We must accept that our usual ideas of space and time do not apply to the very early universe. That is beyond our experience, but not beyond our imagination, or our mathematics. If in the early universe all four dimensions behave like space, what happens to the beginning of time?

The realization that time can behave like another direction of space means one can get rid of the problem of time having a beginning, in a similar way in which we got rid of the edge of the world. Suppose the beginning of the universe was like the South

Pole of the earth, with degrees of latitude playing the role of time. As one moves north, the circles of constant latitude, representing the size of the universe, would expand. The universe would start as a point at the South Pole, but the South Pole is much like any other point. To ask what happened before the beginning of the universe would become a meaningless question, because there is nothing south of the South Pole. In this picture space-time has no boundary—the same laws of nature hold at the South Pole as in other places. In an analogous manner, when one combines the general theory of relativity with quantum theory, the question of what happened before the beginning of the universe is rendered meaningless. This idea that histories should be closed surfaces without boundary is called the no-boundary condition.

Over the centuries many, including Aristotle, believed that the universe must have always existed in order to avoid the issue of how it was set up. Others believed the universe had a beginning, and used it as an argument for the existence of God. The realization that time behaves like space presents a new alternative. It removes the age-old objection to the universe having a beginning, but also means that the beginning of the universe was governed by the laws of science and doesn't need to be set in motion by some god.

If the origin of the universe was a quantum event, it should be accurately described by the Feynman sum over histories. To apply quantum theory to the entire universe—where the observers are part of the system being observed—is tricky, however. In Chapter 4 we saw how particles of matter fired at a screen with two slits in it could exhibit interference patterns just as water waves do. Feynman showed that this arises because a particle does not have a unique history. That is, as it moves from its starting point A to some endpoint B, it doesn't take one definite path, but rather simultaneously takes every possible path connecting the two points. From this point of view, interference is no surprise because, for instance, the particle can travel through both slits at the same time and interfere with itself. Applied to the motion of a particle, Feynman's method tells us that to calculate the probability of any particular endpoint we need to consider all the possible histories that the particle might follow from its starting point to that endpoint. One can also use Feynman's methods to calculate the quantum probabilities for observations of the universe. If they are applied to the universe as a whole, there is no point A, so we add up all the histories that satisfy the no-boundary condition and end at the universe we observe today.

In this view, the universe appeared spontaneously, starting off in every possible way. Most of these correspond to other universes. While some of those universes are similar to ours, most are very different. They aren't just different in details, such as whether Elvis really did die young or whether turnips are a dessert food, but rather they differ even in their apparent laws of nature. In fact, many universes exist with many different sets of physical laws. Some people make a great mystery of this idea, sometimes called the multiverse concept, but these are just different expressions of the Feynman sum over histories.

To picture this, let's alter Eddington's balloon analogy and instead think of the expanding universe as the surface of a bubble. Our picture of the spontaneous quantum creation of the universe is then a bit like the formation of bubbles of steam in boiling water. Many tiny bubbles appear, and then disappear again. These represent mini-universes that expand but collapse again while still of microscopic size. They represent possible alternative universes, but they are not of much interest since they do not last long enough to develop galaxies and stars, let alone intelligent life. A few of the little bubbles, however, will grow large enough so that they will be safe from recollapse. They will continue to expand at an ever-increasing rate and will form the bubbles of steam we are able to see. These correspond to universes that start off expanding at an ever-increasing rate—in other words, universes in a state of inflation.

Multiverse Quantum fluctuations lead to the creation of tiny universes out of nothing. A few of these reach a critical size, then expand in an inflationary manner, forming galaxies, stars and, in at least one case, beings like us.

As we said, the expansion caused by inflation would not be completely uniform. In the sum over histories, there is only one completely uniform and regular history, and it will have the greatest probability, but many other histories that are very slightly irregular will have probabilities that are almost as high. That is why inflation predicts that the early universe is likely to be slightly nonuniform, corresponding to the small variations in the temperature that were observed in the CMBR. The irregularities in the early universe are lucky for us. Why? Homogeneity is good if you don't want cream separating out from your milk, but a uniform universe is a boring universe. The irregularities in the early universe are important because if some regions had a slightly higher density than others, the gravitational attraction of the extra density would slow the expansion of that region compared with its surroundings. As the force of gravity slowly draws matter together, it can eventually cause it to

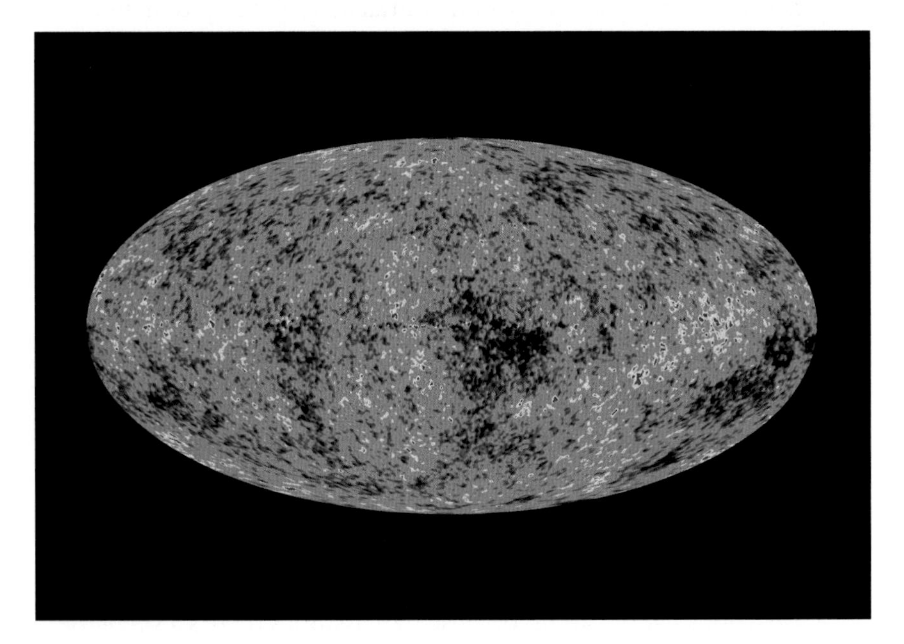

The Microwave Background This picture of the sky was created from seven years of WMAP data released in 2010. It reveals temperature fluctuations—shown as colour differences—dating back 13.7 billion years. The fluctuations pictured correspond to temperature differences of less than a thousandth of a degree on the Centigrade scale. Yet they were the seeds that grew to become the galaxies. Credit: NASA/WMAP Science Team.

collapse to form galaxies and stars, which can lead to planets and, on at least one occasion, people. So look carefully at the map of the microwave sky. It is the blueprint for all the structure in the universe. We are the product of quantum fluctuations in the very early universe. If one were religious, one could say that God really does play dice.

This idea leads to a view of the universe that is profoundly different from the traditional concept, requiring us to adjust the way that we think about the history of the universe. In order to make predictions in cosmology, we need to calculate the probabilities of different states of the entire universe at the present time. In physics one normally assumes some initial state for a system and evolves it forward in time employing the relevant mathematical equations. Given the state of a system at one time, one tries to calculate the probability that the system will be in some different state at a later time. The usual assumption in cosmology is that the universe has a single definite history. One can use the laws of physics to calculate how this history develops with time. We call this the "bottom-up" approach to cosmology. But since we must take into account the quantum nature of the universe as expressed by the Feynman sum over histories, the probability amplitude that the universe is now in a particular state is arrived at by adding up the contributions from all the histories that satisfy the no-boundary condition and end in the state in question. In cosmology, in other words, one shouldn't follow the history of the universe from the bottom up because that assumes there's a single history, with a well-defined starting point and evolution. Instead, one should trace the histories from the top down, backwards from the present time. Some histories will be more probable than others, and the sum will normally be dominated by a single history that starts with the creation of the universe and culminates in the state under consideration. But there will be different histories for different possible states of the universe at the present time. This leads to a radically different view of cosmology, and the relation between cause and effect. The histories that contribute to the Feynman sum don't have an independent existence, but depend on what is being measured. We create history by our observation, rather than history creating us.

The idea that the universe does not have a unique observer-independent history might seem to conflict with certain facts we know. There might be one history in which the moon is made of Roquefort cheese. But we have observed that the moon is not made of cheese, which is bad news for mice. Hence histories in which the moon is made of cheese do not contribute to the present state of our universe, though they might contribute to others. That might sound like science fiction, but it isn't.

An important implication of the top-down approach is that the apparent laws of nature depend on the history of the universe. Many scientists believe there exists a single theory that explains those laws as well as nature's physical constants, such as the mass of the electron or the dimensionality of space-time. But top-down cosmology dictates that the apparent laws of nature are different for different histories.

Consider the apparent dimension of the universe. According to M-theory, space-time has ten space dimensions and one time dimension. The idea is that seven of the space dimensions are curled up so small that we don't notice them, leaving us with the illusion that all that exist are the three remaining large dimensions we are familiar with. One of the central open questions in M-theory is: why, in our universe, aren't there more large dimensions, and why are any dimensions curled up?

Many people would like to believe that there is some mechanism that causes all but three of the space dimensions to curl up spontaneously. Alternatively, maybe all dimensions started small, but for some understandable reason three space dimensions expanded and the rest did not. It seems, however, that there is no dynamical reason for the universe to appear four-dimensional. Instead, top-down cosmology predicts that the number of large space dimensions is not fixed by any principle of physics. There will be a quantum probability amplitude for every number of large space dimensions from zero to ten. The Feynman sum allows for all of these, for every possible history for the universe, but the observation that our universe has three large space dimensions selects out the subclass of histories that have the property that is being observed. In other words, the quantum probability that the universe has more or less than three large space dimensions is irrelevant because we have already determined that we are in a universe with three large space dimensions. So as long as the probability amplitude for three large space dimensions is not exactly zero, it doesn't matter how small it is compared with the probability amplitude for other numbers of dimensions. It would be like asking for the probability amplitude that the present pope is Chinese. We know that he is German, even though the probability that he is Chinese is higher because there are more Chinese than there are Germans. Similarly, we know our universe exhibits three large space dimensions, and so even though other numbers of large space dimensions may have a greater probability amplitude, we are interested only in histories with three.

What about the curled-up dimensions? Recall that in M-theory the precise shape of the remaining curled-up dimensions, the internal space, determines both the values of physical quantities such as the charge on the electron and the nature of the interactions between elementary particles, that is, the forces of nature. Things would have worked out neatly if M-theory had allowed just one shape for the curled dimensions, or perhaps a few, all but one of which might have been ruled out by some means, leaving us with just one possibility for the apparent laws of nature. Instead, there are probability amplitudes for perhaps as many as 10^{500} different internal spaces, each leading to different laws and values for the physical constants.

If one builds the history of the universe from the bottom up, there is no reason the universe should end up with the internal space for the particle interactions that we actually observe, the standard model (of elementary particle interactions). But in the top-down approach we accept that universes exist with all possible internal spaces. In some universes electrons have the weight of golf balls and the force of gravity is stronger than that of magnetism. In ours, the standard model, with all its parameters, applies. One can calculate the probability amplitude for the internal space that leads to the standard model on the basis of the no-boundary condition. As with the probability of there being a universe with three large space dimensions, it doesn't matter how small this amplitude is relative to other possibilities because we already observed that the standard model describes our universe.

The theory we describe in this chapter is testable. In the prior examples we emphasized that the relative probability amplitudes for radically different universes, such as those with a different number of large space dimensions, don't matter. The relative probability amplitudes for neighbouring (i.e. similar) universes, however, are important. The no-boundary condition implies that the probability amplitude is highest for histories in which the universe starts out completely smooth. The amplitude is reduced for universes that are more irregular. This means that the early universe would have been almost smooth, but with small irregularities. As we've noted, we can observe these irregularities as small variations in the microwaves coming from different directions in the sky. They have been found to agree exactly with the general demands of inflation theory; however, more precise measurements are needed to fully differentiate the top-down theory from others, and to either support or refute it. These may well be carried out by satellites in the future.

Hundreds of years ago people thought the earth was unique, and situated at the centre of the universe. Today we know there are hundreds of billions of stars in our galaxy, a large percentage of them with planetary systems, and hundreds of billions of galaxies. The results described in this chapter indicate that our universe itself is also one of many, and that its apparent laws are not uniquely determined. This must be disappointing for those who hoped that an ultimate theory, a theory of everything, would predict the nature of everyday physics. We cannot predict discrete features such as the number of large space dimensions or the internal space that determines the physical quantities we observe (e.g. the mass and charge of the electron and other elementary particles). Rather, we use those numbers to select which histories contribute to the Feynman sum.

We seem to be at a critical point in the history of science, in which we must alter our conception of goals and of what makes a physical theory acceptable. It appears that the fundamental numbers, and even the form, of the apparent laws of nature are not demanded by logic or physical principle. The parameters are free to take on many values and the laws to take on any form that leads to a self-consistent mathematical theory, and they do take on different values and different forms in different universes. That may not satisfy our human desire to be special or to discover a neat package to contain all the laws of physics, but it does seem to be the way of nature.

There seems to be a vast landscape of possible universes. However, as we'll see in the next chapter, universes in which life like us can exist are rare. We live in one in which life is possible, but if the universe were only slightly different, beings like us could not exist. What are we to make of this fine-tuning? Is it evidence that the universe, after all, was designed by a benevolent creator? Or does science offer another explanation?

7

THE APPARENT MIRACLE

THE CHINESE TELL OF A TIME during the Hsia dynasty (c. 2205—c. 1782 BC) when our cosmic environment suddenly changed. Ten suns appeared in the sky. The people on earth suffered greatly from the heat, so the emperor ordered a famous archer to shoot down the extra suns. The archer was rewarded with a pill that had the power to make him immortal, but his wife stole it. For that offence she was banished to the moon.

The Chinese were right to think that a solar system with ten suns is not friendly to human life. Today we know that, while perhaps offering great tanning opportunities, any solar system with multiple suns would probably never allow life to develop. The reasons are not quite as simple as the searing heat imagined in the Chinese legend. In fact, a planet could experience a pleasant temperature while orbiting multiple stars, at least for a while. But uniform heating over long periods of time, a situation that seems necessary for life, would be unlikely. To understand why, let's look at what happens in the simplest type of multiple-star system, one with two suns, which is called a binary system. About half of all stars in the sky are members of such systems. But even simple binary systems can maintain only certain kinds of stable orbits, of the type shown below. In each of these orbits there would likely be a time in which the planet would be either too hot or too cold to sustain life. The situation is even worse for clusters having many stars.

Our solar system has other "lucky" properties without which sophisticated life-forms might never have evolved. For example, Newton's laws allow for planetary orbits to be either circles or

Binary Orbits Planets that orbit binary star systems will probably have inhospitable weather, in some seasons too hot for life, in others, too cold.

ellipses (ellipses are squashed circles, wider along one axis and narrower along another). The degree to which an ellipse is squashed is described by what is called its eccentricity, a number between zero and one. An eccentricity near zero means the figure resembles a circle, whereas an eccentricity near one means it is very flattened. Kepler was upset by the idea that planets don't move in perfect circles, but the earth's orbit has an eccentricity of only about 2 percent, which means it is nearly circular. As it turns out, that is a stroke of very good fortune.

Seasonal weather patterns on earth are determined mainly by the tilt of the earth's axis of rotation relative to the plane of its orbit around the sun. During winter in the Northern Hemisphere, for example, the North Pole is tilted away from the sun. The fact that the earth is closest to the sun at that time—only 91.5 million miles

Eccentricities Eccentricity is a measure of how near an ellipse is to a circle. Circular orbits are friendly to life, while very elongated orbits result in large seasonal temperature fluctuations.

away, as opposed to around 94.5 million miles away from the sun in early July—has a negligible effect on the temperature compared with the effect of its tilt. But on planets with a large orbital eccentricity, the varying distance from the sun plays a much larger role. On Mercury, for example, with a 20 percent eccentricity, the temperature is over 200 degrees Fahrenheit warmer at the planet's closest approach to the sun (perihelion) than when it is at its farthest from the sun (aphelion). In fact, if the eccentricity of the earth's orbit were near one, our oceans would boil when we reached our nearest point to the sun, and freeze over when we reached our farthest, making neither winter nor summer vacations very pleasant. Large orbital eccentricities are not conducive to life, so we are fortunate to have a planet for which orbital eccentricity is near zero.

We are also lucky in the relationship of our sun's mass to our distance from it. That is because a star's mass determines the amount of energy it gives off. The largest stars have a mass about a hundred times that of our sun, while the smallest are about a hundred times less massive. And yet, assuming the earth-sun distance as a given, if our sun were just 20 percent less or more massive, the earth would be colder than present-day Mars or hotter than present-day Venus.

Traditionally, given any star, scientists define the habitable zone as the narrow region around the star in which temperatures are

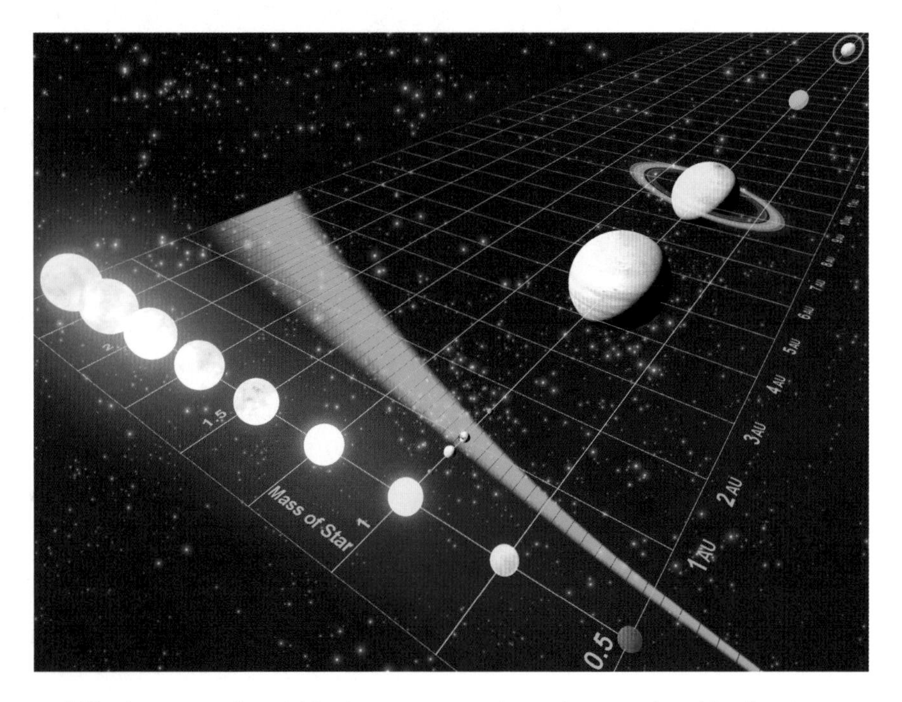

Goldilocks Zone If Goldilocks were sampling planets, she'd find only those within the green zone suitable for life. The yellow star represents our own sun. The whiter stars are larger and hotter, the redder ones smaller and cooler. Planets closer to their suns than the green zone would be too hot for life, and planets beyond it too cold. The size of the hospitable zone is smaller for cooler stars.

such that liquid water can exist. The habitable zone is sometimes called the "Goldilocks zone", because the requirement that liquid water exist means that, like Goldilocks, the development of intelligent life requires that planetary temperatures be "just right". The habitable zone in our solar system, pictured above, is tiny. Fortunately for those of us who are intelligent life-forms, the earth fell within it!

Newton believed that our strangely habitable solar system did not "arise out of chaos by the mere laws of nature". Instead, he maintained, the order in the universe was "created by God at first and conserved by him to this Day in the same state and condition". It is easy to understand why one might think that. The many improbable occurrences that conspired to enable our existence, and our world's human-friendly design, would indeed be puzzling if ours were the only solar system in the universe. But in 1992 came the first confirmed observation of a planet orbiting a star other than our sun. We now know of hundreds of such planets, and few doubt that there exist countless others among the many billions of stars in our universe. That makes the coincidences of our planetary conditions—the single sun, the lucky combination of earth-sun distance and solar mass—far less remarkable, and far less compelling as evidence that the earth was carefully designed just to please us human beings. Planets of all sorts exist. Some—or at least one—support life. Obviously, when the beings on a planet that supports life examine the world around them, they are bound to find that their environment satisfies the conditions they require to exist.

It is possible to turn that last statement into a scientific principle: our very existence imposes rules determining from where and at what time it is possible for us to observe the universe. That is, the fact of our being restricts the characteristics of the kind of environment in which we find ourselves. That principle is called the weak anthropic principle. (We'll see shortly why the adjective "weak" is attached.) A better term than "anthropic principle" would have been "selection principle", because the principle refers to how our own knowledge of our existence imposes rules that select, out of all the possible environments, only those environments with the characteristics that allow life.

Though it may sound like philosophy, the weak anthropic principle can be used to make scientific predictions. For example, how old is the universe? As we'll soon see, for us to exist the universe must contain elements such as carbon, which are produced by cooking lighter elements inside stars. The carbon must then be scattered through space in a supernova explosion, and eventually condense as part of a planet in a new-generation solar system. In 1961 physicist Robert Dicke argued that the process takes about 10 billion years, so our being here means that the universe must be at least that old. On the other hand, the universe cannot be much older than 10 billion years, since in the far future all the fuel for stars will have been used up, and we require hot stars for our sustenance. Hence the universe must be about 10 billion years old. That is not an extremely precise prediction, but it is true—according to current data the big bang occurred about 13.7 billion years ago.

As was the case with the age of the universe, anthropic predictions usually produce a range of values for a given physical parameter rather than pinpointing it precisely. That's because our existence, while it might not require a particular value of some physical parameter, often is dependent on such parameters not varying too far from where we actually find them. We furthermore expect that the actual conditions in our world are typical within the anthropically allowed range. For example, if only modest orbital eccentricities, say between zero and 0.5, will allow life, then an eccentricity of 0.1 should not surprise us because among all the planets in the universe, a fair percentage probably have orbits with eccentricities that small. But if it turned out that the earth moved in a near-perfect circle, with eccentricity, say, of 0.00000000001, that would make the earth a very special planet indeed, and motivate us to try to explain why we find ourselves living in such an anomalous home. That idea is sometimes called the principle of mediocrity.

The lucky coincidences pertaining to the shape of planetary orbits, the mass of the sun, and so on are called environmental because they arise from the serendipity of our surroundings and not from a fluke in the fundamental laws of nature. The age of the universe is also an environmental factor, since there are an earlier and a later time in the history of the universe, but we must live in this era because it is the only era conducive to life. Environmental coincidences are easy to understand because ours is only one cosmic habitat among many that exist in the universe, and we obviously must exist in a habitat that supports life.

The weak anthropic principle is not very controversial. But there is a stronger form that we will argue for here, although it is regarded with disdain among some physicists. The strong anthropic principle suggests that the fact that we exist imposes constraints not just on our *environment* but on the possible *form and content of the laws of nature* themselves.The idea arose because it is not only the peculiar characteristics of our solar system that seem oddly conducive to the development of human life but also the characteristics of our entire universe, and that is much more difficult to explain.

The tale of how the primordial universe of hydrogen, helium, and a bit of lithium evolved to a universe harbouring at least one world with intelligent life like us is a tale of many chapters. As we mentioned earlier, the forces of nature had to be such that heavier elements—especially carbon—could be produced from the primordial elements, and remain stable for at least billions of years. Those heavy elements were formed in the furnaces we call stars, so the forces first had to allow stars and galaxies to form. Those grew from the seeds of tiny inhomogeneities in the early universe, which was almost completely uniform but thankfully contained density variations of about 1 part in 100,000. However, the existence of stars, and the existence inside those stars of the elements we are made of, is not enough. The dynamics of the stars had to be such that some would eventually explode, and, moreover, explode precisely in a way that could disburse the heavier elements through space. In addition, the laws of nature had to dictate that those remnants could recondense into a new generation of stars, these surrounded by planets incorporating the newly formed heavy elements. Just as certain events on early earth had to occur in order to allow us to develop, so too was each link of this chain necessary for our existence. But in the case of the events resulting in the evolution of the universe, such developments were governed by the balance of the fundamental forces of nature, and it is those whose interplay had to be just right in order for us to exist.

One of the first to recognize that this might involve a good measure of serendipity was Fred Hoyle, in the 1950s. Hoyle believed that all chemical elements had originally been formed from hydrogen, which he felt was the true primordial substance. Hydrogen has the simplest atomic nucleus, consisting of just one proton, either alone or in combination with one or two neutrons. (Different forms of hydrogen, or any nucleus, having the same number of protons but different numbers of neutrons are called isotopes.) Today we know that helium and lithium, atoms whose nuclei contain two and three protons, were also primordially synthesized, in much smaller amounts, when the universe was about 200 seconds old. Life, on the other hand, depends on more complex elements. Carbon is the most important of these, the basis for all organic chemistry.

Though one might imagine "living" organisms such as intelligent computers produced from other elements, such as silicon, it is doubtful that life could have *spontaneously* evolved in the absence of carbon. The reasons for that are technical but have to do with the unique manner in which carbon bonds with other elements. Carbon dioxide, for example, is gaseous at room temperature, and biologically very useful. Since silicon is the element directly below carbon on the periodic table, it has similar chemical properties. However, silicon dioxide, quartz, is far more useful in a rock collection than in an organism's lungs. Still, perhaps lifeforms could evolve that feast on silicon and rhythmically twirl their tails in pools of liquid ammonia. Even that type of exotic life could not evolve from just the primordial elements, for those elements can form only two stable compounds, lithium hydride, which is a colourless crystalline solid, and hydrogen gas, neither of them a compound likely to reproduce or even to fall in love. Also, the fact remains that *we* are a carbon life-form, and that raises the issue of how carbon, whose nucleus contains six protons, and the other heavy elements in our bodies were created.

The first step occurs when older stars start to accumulate helium, which is produced when two hydrogen nuclei collide and fuse with each other. This fusion is how stars create the energy that warms us. Two helium atoms can in turn collide to form beryllium, an atom whose nucleus contains four protons. Once beryllium is formed, it could in principle fuse with a third helium nucleus to form carbon. But that doesn't happen, because the isotope of beryllium that is formed decays almost immediately back into helium nuclei.

The situation changes when a star starts to run out of hydrogen. When that happens the star's core collapses until its central temperature rises to about 100 million degrees Kelvin. Under those conditions, nuclei encounter each other so often that some beryllium nuclei collide with a helium nucleus before they have had a chance to decay. Beryllium can then fuse with helium to form an isotope of carbon that is stable. But that carbon is still a long way from forming ordered aggregates of chemical compounds of the type that can enjoy a glass of Bordeaux, juggle flaming bowling pins or ask questions about the universe. For beings such as humans to exist, the carbon must be moved from inside the star to friendlier neighbourhoods. That, as we've said, is ac-

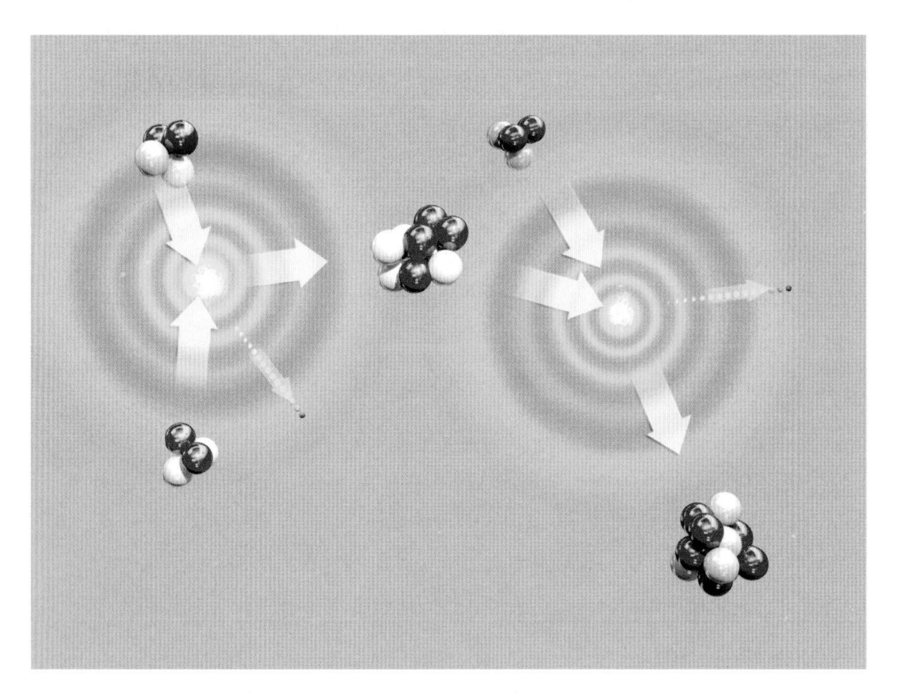

Triple Alpha Process Carbon is made inside stars from the collisions of three helium nuclei, an event that would be very unlikely if not for a special property of the laws of nuclear physics.

complished when the star, at the end of its life cycle, explodes as a supernova, expelling carbon and other heavy elements that later condense into a planet.

This process of carbon creation is called the triple alpha process because "alpha particle" is another name for the nucleus of the isotope of helium involved, and because the process requires that three of them (eventually) fuse together. The usual physics predicts that the rate of carbon production via the triple alpha process ought to be quite small. Noting this, in 1952 Hoyle predicted that the sum of the energies of a beryllium nucleus and a helium nucleus must be almost exactly the energy of a certain quantum state of the isotope of carbon formed, a situation called a resonance, which greatly increases the rate of a nuclear reaction. At the time, no such energy level was known, but based on Hoyle's suggestion, William Fowler at Caltech sought and found it, providing important support for Hoyle's views on how complex nuclei were created.

Hoyle wrote, "I do not believe that any scientist who examined the evidence would fail to draw the inference that the laws of nuclear physics have been deliberately designed with regard to the consequences they produce inside the stars." At the time no one knew enough nuclear physics to understand the magnitude of the serendipity that resulted in these exact physical laws. But in investigating the validity of the strong anthropic principle, in recent years physicists began asking themselves what the universe would have been like if the laws of nature were different. Today we can create computer models that tell us how the rate of the triple alpha reaction depends upon the strength of the fundamental forces of nature. Such calculations show that a change of as little as 0.5 percent in the strength of the strong nuclear force, or 4 percent in the electric force, would destroy either nearly all carbon or all oxygen in every star, and hence the possibility of life as we know it. Change those rules of our universe just a bit, and the conditions for our existence disappear!

By examining the model universes we generate when the theories of physics are altered in certain ways, one can study the effect of changes to physical law in a methodical manner. It turns out that it is not only the strengths of the strong nuclear force and the electromagnetic force that are made to order for our existence. Most of the fundamental constants in our theories appear fine-tuned in the sense that if they were altered by only modest amounts, the universe would be qualitatively different, and in many cases unsuitable for the development of life. For example, if the other nuclear force, the weak force, were much weaker, in the early universe all the hydrogen in the cosmos would have turned to helium, and hence there would be no normal stars; if it were much stronger, exploding supernovas would not eject their outer envelopes, and hence would fail to seed interstellar space with the heavy elements planets require to foster life. If protons were 0.2 percent heavier, they would decay into neutrons, destabilizing atoms. If the sum of the masses of the types of quark that make up a proton were changed by as little as 10 percent, there would be far fewer of the stable atomic nuclei of which we are made; in fact, the summed quark masses seem roughly optimized for the existence of the largest number of stable nuclei.

If one assumes that a few hundred million years in stable orbit are necessary for planetary life to evolve, the number of space dimensions is also fixed by our existence. That is because, according to the laws of gravity, it is only in three dimensions that stable elliptical orbits are possible. Circular orbits are possible in other dimensions, but those, as Newton feared, are unstable. In any but three dimensions even a small disturbance, such as that produced by the pull of the other planets, would send a planet off its circular orbit and cause it to spiral either into or away from the sun, so we would either burn up or freeze. Also, in more than three dimensions the gravitational force between two bodies would decrease more rapidly than it does in three dimensions. In three dimensions the gravitational force drops to ¼ of its value if one doubles the distance. In four dimensions it would drop to ⅛, in five dimensions it would drop to 1/16, and so on. As a result, in more than three dimensions the sun would not be able to exist in a stable state with its internal pressure balancing the pull of gravity. It would either fall apart or collapse to form a black hole, either of which could ruin your day. On the atomic scale, the electrical forces would behave in the same way as gravitational forces. That means the electrons in atoms would either escape or spiral into the nucleus. In neither case would atoms as we know them be possible.

The emergence of the complex structures capable of supporting intelligent observers seems to be very fragile. The laws of nature form a system that is extremely fine-tuned, and very little in physical law can be altered without destroying the possibility of the development of life as we know it. Were it not for a series of startling coincidences in the precise details of physical law, it seems, humans and similar life-forms would never have come into being.

The most impressive fine-tuning coincidence involves the so-called cosmological constant in Einstein's equations of general relativity. As we've said, in 1915, when he formulated the theory, Einstein believed that the universe was static, that is, neither expanding nor contracting. Since all matter attracts other matter, he introduced into his theory a new antigravity force to combat the tendency of the universe to collapse onto itself. This force, unlike other forces, did not come from any particular source but was built into the very fabric of space-time. The cosmological constant describes the strength of that force.

When it was discovered that the universe was not static, Einstein eliminated the cosmological constant from his theory and called including it the greatest blunder of his life. But in 1998 observations of very distant supernovas revealed that the universe is expanding at an accelerating rate, an effect that is not possible without some kind of repulsive force acting throughout space. The cosmological constant was resurrected. Since we now know that its value is not zero, the question remains, why does it have the value that it does? Physicists have created arguments explaining how it might arise due to quantum mechanical effects, but the value they calculate is about 120 orders of magnitude (a 1 followed by 120 zeroes) stronger than the actual value, obtained through the supernova observations. That means that either the reasoning employed in the calculation was wrong or else some other effect exists that miraculously cancels all but an unimaginably tiny fraction of the number calculated. The one thing that is certain is that if the value of the cosmological constant were much larger than it is, our universe would have blown itself apart before galaxies could form and—once again—life as we know it would be impossible.

What can we make of these coincidences? Luck in the precise form and nature of fundamental physical law is a different kind of luck from the luck we find in environmental factors. It cannot be so easily explained, and has far deeper physical and philosophical implications. Our universe and its laws appear to have a design that both is tailor-made to support us and, if we are to exist, leaves little room for alteration. That is not easily explained, and raises the natural question of why it is that way.

Many people would like us to use these coincidences as evidence of the work of God. The idea that the universe was designed to accommodate mankind appears in theologies and mythologies dating from thousands of years ago right up to the present. In the Mayan Popol Vuh mythohistorical narratives the gods proclaim, "We shall receive neither glory nor honour from all that we have created and formed until human beings exist, endowed with sentience." A typical Egyptian text dated 2000 BC states, "Men, the cattle of God, have been well provided for. He [the sun god] made the sky and earth for their benefit." In China the Taoist philosopher Lieh Yü-K'ou (c. 400 BC) expressed the idea through a character in a tale who says, "Heaven makes the five kinds of grain to grow, and brings forth the finny and the feathered tribes, especially for our benefit."

In Western culture the Old Testament contains the idea of providential design in its story of creation, but the traditional Christian viewpoint was also greatly influenced by Aristotle, who believed "in an intelligent natural world that functions according to some deliberate design". The medieval Christian theologian Thomas Aquinas employed Aristotle's ideas about the order in nature to argue for the existence of God. In the eighteenth century another Christian theologian went so far as to say that rabbits have white tails in order that it be easy for us to shoot them. A more modern illustration of the Christian view was given a few years ago when Cardinal Christoph Schönborn, archbishop of Vienna, wrote, "Now, at the beginning of the 21st century, faced with scientific claims like neo-Darwinism and the multiverse [many universes] hypothesis in cosmology invented to avoid the overwhelming evidence for purpose and design found in modern science, the Catholic Church will again defend human nature by proclaiming that the immanent design in nature is real." In cosmology the overwhelming evidence for purpose and design to which the cardinal was referring is the fine-tuning of physical law we described above.

The turning point in the scientific rejection of a human-centred universe was the Copernican model of the solar system, in which the earth no longer held a central position. Ironically, Copernicus's own worldview was anthropomorphic, even to the extent that he comforts us by pointing out that, despite his heliocentric model, the earth is *almost* at the universe's centre: "Although [the earth] is not at the centre of the world, nevertheless the distance [to that centre] is as nothing in particular when compared to that of the fixed stars." With the invention of the telescope, observations in the seventeenth century, such as the fact that ours is not the only planet orbited by a moon, lent weight to the principle that we hold no privileged position in the universe. In the ensuing centuries the more we discovered about the universe, the more it seemed ours was probably just a garden-variety planet. But the discovery relatively recently of the extreme fine-tuning of so many of the laws of nature could lead at least some of us some back to the old idea that this grand design is the work of some grand designer. In the United States, because the Constitution prohibits the teaching of religion in schools, that type of idea is called intelligent design, with the unstated but implied understanding that the designer is God.

That is not the answer of modern science. We saw in Chapter 5 that our universe seems to be one of many, each with different laws. That multiverse idea is not a notion invented to account for the miracle of fine-tuning. It is a consequence of the no-boundary condition as well as many other theories of modern cosmology. But if it is true, then the strong anthropic principle can be considered effectively equivalent to the weak one, putting the fine-tunings of physical law on the same footing as the environmental factors, for it means that our cosmic habitat—now the entire observable universe—is only one of many, just as our solar system is one of many. That means that in the same way that the environmental coincidences of our solar system were rendered unremarkable by the realization that billions of such systems exist, the fine-tunings in the laws of nature can be explained by the existence of multiple universes. Many people through the ages have attributed to God the beauty and complexity of nature that in their time seemed to have no scientific explanation. But just as Darwin and Wallace explained how the apparently miraculous design of living forms could appear without intervention by a supreme being, the multiverse concept can explain the fine-tuning of physical law without the need for a benevolent creator who made the universe for our benefit.

Einstein once posed to his assistant Ernst Straus the question "Did God have any choice when he created the universe?" In the late sixteenth century Kepler was convinced that God had created the universe according to some perfect mathematical principle. Newton showed that the same laws that apply in the heavens apply on earth, and developed mathematical equations to express those laws that were so elegant they inspired almost religious fervour among many eighteenth-century scientists, who seemed intent on using them to show that God was a mathematician.

Ever since Newton, and especially since Einstein, the goal of physics has been to find simple mathematical principles of the kind Kepler envisioned, and with them to create a unified theory of everything that would account for every detail of the matter and forces we observe in nature. In the late nineteenth and early twentieth century Maxwell and Einstein united the theories of electricity, magnetism and light. In the 1970s the standard model was created, a single theory of the strong and weak nuclear forces, and the electromagnetic force. String theory and M-theory then came into being in an attempt to include the remaining force, gravity. The goal was to find not just a single theory that explains all the forces but also one that explains the fundamental numbers we have been talking about, such as the strength of the forces and the masses and charges of the elementary particles. As Einstein put it, the hope was to be able to say that "nature is so constituted that it is possible logically to lay down such strongly determined laws that within these laws only rationally completely determined constants occur (not constants, therefore, whose numerical value could be changed without destroying the theory)". A unique theory would be unlikely to have the fine-tuning that allows us to exist. But if in light of recent advances we interpret Einstein's dream to be that of a unique theory that explains this and other universes, with their whole spectrum of different laws, then M-theory could be that theory. But is M-theory unique, or demanded by any simple logical principle? Can we answer the question, *why M-theory*?

8

THE GRAND DESIGN

IN THIS BOOK WE HAVE DESCRIBED how regularities in the motion of astronomical bodies such as the sun, the moon and the planets suggested that they were governed by fixed laws rather than being subject to the arbitrary whims and caprices of gods and demons. At first the existence of such laws became apparent only in astronomy (or astrology, which was regarded as much the same). The behaviour of things on earth is so complicated and subject to so many influences that early civilizations were unable to discern any clear patterns or laws governing these phenomena. Gradually, however, new laws were discovered in areas other than astronomy, and this led to the idea of scientific determinism: there must be a complete set of laws that, given the state of the universe at a specific time, would specify how the universe would develop from that time forward. These laws should hold everywhere and at all times; otherwise they wouldn't be laws. There could be no exceptions or miracles. Gods or demons couldn't intervene in the running of the universe.

At the time that scientific determinism was first proposed, Newton's laws of motion and gravity were the only laws known. We have described how these laws were extended by Einstein in his general theory of relativity, and how other laws were discovered to govern other aspects of the universe.

The laws of nature tell us *how* the universe behaves, but they don't answer the *why*? questions that we posed at the start of this book:

Why is there something rather than nothing?
Why do we exist?
Why this particular set of laws and not some other?

Some would claim the answer to these questions is that there is a God who chose to create the universe that way. It is reasonable to ask who or what created the universe, but if the answer is God, then the question has merely been deflected to that of who created God. In this view it is accepted that some entity exists that needs no creator, and that entity is called God. This is known as the first-cause argument for the existence of God. We claim, however, that it is possible to answer these questions purely within the realm of science, and without invoking any divine beings.

According to the idea of model-dependent realism introduced in Chapter 3, our brains interpret the input from our sensory organs by making a model of the outside world. We form mental concepts of our home, trees, other people, the electricity that flows from wall sockets, atoms, molecules and other universes. These mental concepts are the only reality we can know. There is no model-independent test of reality. It follows that a well-constructed model creates a reality of its own. An example that can help us think about issues of reality and creation is the Game of Life, invented in 1970 by a young mathematician at Cambridge named John Conway.

The word "game" in the Game of Life is a misleading term. There are no winners and losers; in fact, there are no players. The Game of Life is not really a game but a set of laws that govern a two-dimensional universe. It is a deterministic universe: once you set up a starting configuration, or initial condition, the laws determine what happens in the future.

The world Conway envisioned is a square array, like a chessboard, but extending infinitely in all directions. Each square can be in one of two states: alive (shown on page 174 in green) or dead (shown in black). Each square has eight neighbours: the up, down, left and right neighbours and four diagonal neighbours. Time in this world is not continuous but moves forward in discrete steps. Given any arrangement of dead and live squares, the number of live neighbours determine what happens next according to the following laws:

1. A live square with two or three live neighbours survives (survival).
2. A dead square with exactly three live neighbours becomes a live cell (birth).
3. In all other cases a cell dies or remains dead. In the case that a live square has zero or one neighbour, it is said to die of loneliness; if it has more than three neighbours, it is said to die of overcrowding.

That's all there is to it: given any initial condition, these laws generate generation after generation. An isolated living square or two adjacent live squares die in the next generation because they don't have enough neighbours. Three live squares along a diagonal live a bit longer. After the first time step the end squares die, leaving just the middle square, which dies in the following generation. Any diagonal line of squares "evaporates" in just this manner. But if three live squares are placed horizontally in a row, again the centre has two neighbours and survives while the two end squares die, but in this case the cells just above and below the centre cell experience a birth. The row therefore turns into a column. Similarly, the next generation the column back turns into a row, and so forth. Such oscillating configurations are called blinkers.

If three live squares are placed in the shape of an L, a new behaviour occurs. In the next generation the square cradled by the L will give birth, leading to a 2 × 2 block. The block belongs to a

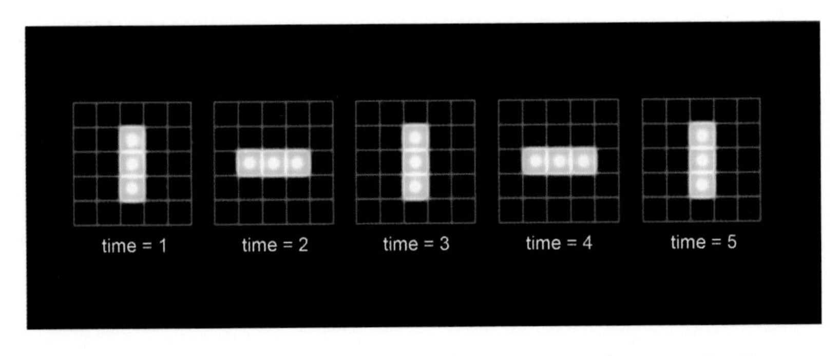

Blinkers Blinkers are a simple type of composite object in the Game of Life.

pattern type called the still life because it will pass from generation to generation unaltered. Many types of patterns exist that morph in the early generations but soon turn into a still life, or die or return to their original form and then repeat the process.

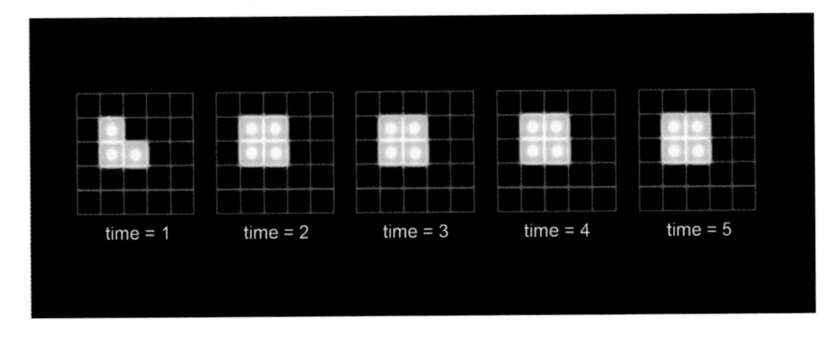

Evolution to a Still Life Some composite objects in the Game of Life evolve into a form that the rules dictate will never change.

There are also patterns called gliders, which morph into other shapes and, after a few generations, return to their original form, but in a position one square down along the diagonal. If you watch these develop over time, they appear to crawl along the array. When these gliders collide, curious behaviours can occur, depending on each glider's shape at the moment of collision.

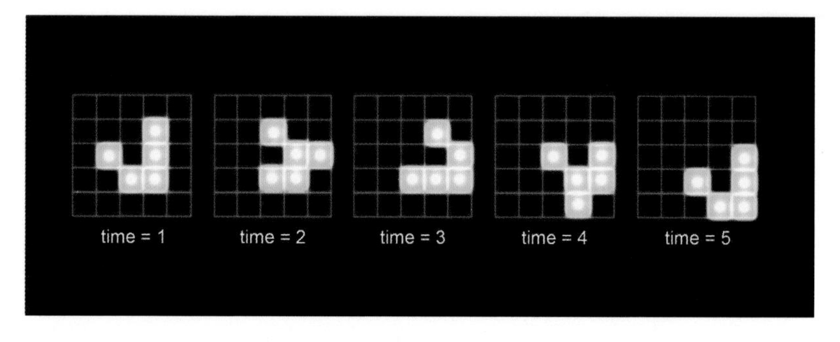

Gliders Gliders morph through these intermediate shapes, then return to their original form, displaced by one square along the diagonal.

What makes this universe interesting is that although the fundamental "physics" of this universe is simple, the "chemistry" can be complicated. That is, composite objects exist on different scales. At the smallest scale, the fundamental physics tells us that there are just live and dead squares. On a larger scale, there are gliders, blinkers and still-life blocks. At a still larger scale there are even more complex objects, such as glider guns: stationary patterns that periodically give birth to new gliders that leave the nest and stream down the diagonal.

If you observed the Game of Life universe for a while on any particular scale, you could deduce laws governing the objects on that scale. For example, on the scale of objects just a few squares across you might have laws such as "Blocks never move", "Gliders move diagonally", and various laws for what happens when objects collide. You could create an entire physics on any level of composite objects. The laws would entail entities and concepts that have no place among the original laws. For example, there are no concepts such as "collide" or "move" in the original laws. Those describe merely the life and death of individual stationary squares. As in our universe, in the Game of Life your reality depends on the model you employ.

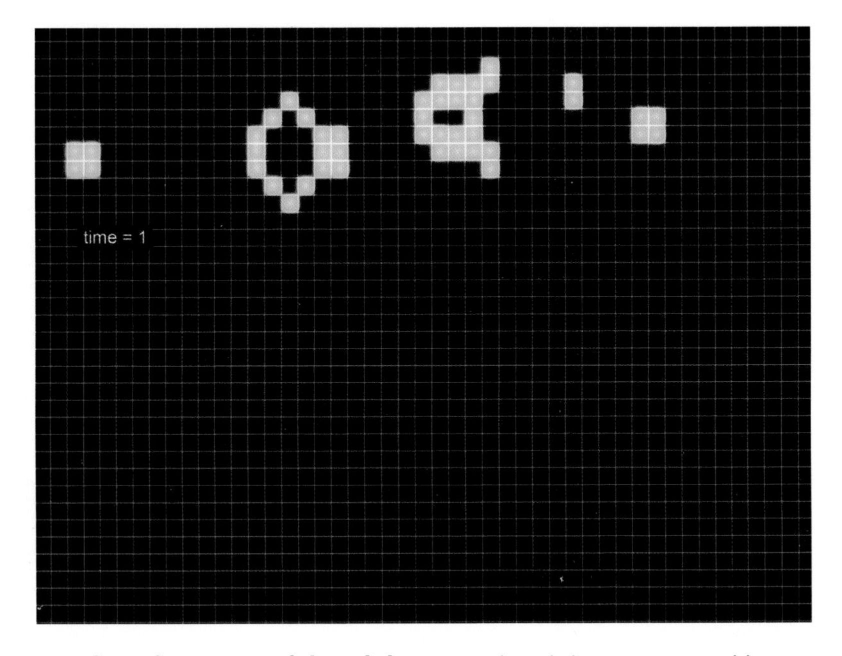

Initial Configuration of the Glider Gun The glider gun is roughly ten times as large as a glider.

Conway and his students created this world because they wanted to know if a universe with fundamental rules as simple as the ones they defined could contain objects complex enough to replicate. In the Game of Life world, do composite objects exist that, after merely following the laws of that world for some generations, will spawn others of their kind? Not only were Conway and his students able to demonstrate that this is possible, but they even showed that such an object would be, in a sense, intelligent! What do we mean by that? To be precise, they showed that the huge conglomerations of squares that self-replicate are "universal Turing machines". For our purposes that means that for any calculation a computer in our physical world can in principle carry out, if the machine were fed the appropriate input—that is, supplied the appropriate Game of Life world environment—then

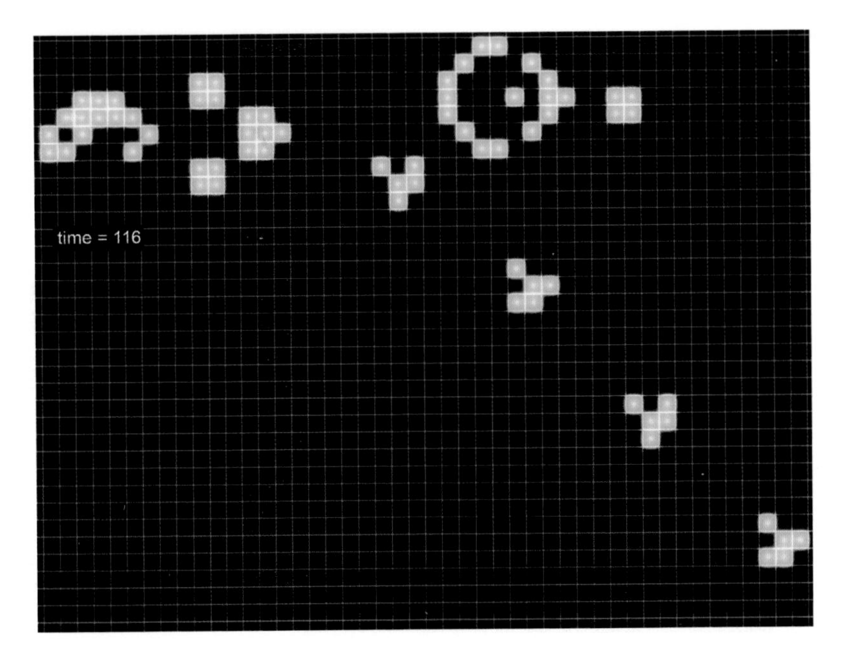

The Glider Gun After 116 Generations With time, the glider gun changes shape, emits a glider and then returns to its original form and position. It repeats the process ad infinitum.

some generations later the machine would be in a state from which an output could be read that would correspond to the result of that computer calculation.

To get a taste for how that works, consider what happens when gliders are shot at a simple 2 × 2 block of live squares. If the gliders approach in just the right way, the block, which had been stationary, will move towards or away from the source of the gliders. In this way, the block can simulate a computer memory. In fact, all the basic functions of a modern computer, such as AND and OR gates, can also be created from gliders. In this manner, just as electrical signals are employed in a physical computer, streams of gliders can be employed to send and process information.

In the Game of Life, as in our world, self-reproducing patterns are complex objects. One estimate, based on the earlier work of mathematician John von Neumann, places the minimum size of a self-replicating pattern in the Game of Life at ten trillion squares—roughly the number of molecules in a single human cell.

One can define living beings as complex systems of limited size that are stable and that reproduce themselves. The objects described above satisfy the reproduction condition but are probably not stable: a small disturbance from outside would probably wreck the delicate mechanism. However, it is easy to imagine that slightly more complicated laws would allow complex systems with all the attributes of life. Imagine a entity of that type, an object in a Conway-type world. Such an object would respond to environmental stimuli, and hence appear to make decisions. Would such life be aware of itself? Would it be self-conscious? This is a question on which opinion is sharply divided. Some people claim that self-awareness is something unique to humans. It gives them free will, the ability to choose between different courses of action.

How can one tell if a being has free will? If one encounters an alien, how can one tell if it is just a robot or it has a mind of its own? The behaviour of a robot would be completely determined, unlike that of a being with free will. Thus one could in principle detect a robot as a being whose actions can be predicted. As we said in Chapter 2, this may be impossibly difficult if the being is large and complex. We cannot even solve exactly the equations for three or more particles interacting with each other. Since an alien the size of a human would contain about a thousand trillion trillion particles even if the alien were a robot, it would be impossible to solve the equations and predict what it would do. We would therefore have to say that any complex being has free will—not as a fundamental feature, but as an effective theory, an admission of our inability to do the calculations that would enable us to predict its actions.

The example of Conway's Game of Life shows that even a very simple set of laws can produce complex features similar to those of intelligent life. There must be many sets of laws with this property. What picks out the fundamental laws (as opposed to the apparent laws) that govern our universe? As in Conway's universe, the laws of our universe determine the evolution of the system, given the state at any one time. In Conway's world we are the creators—we choose the initial state of the universe by specifying objects and their positions at the start of the game.

In a physical universe, the counterparts of objects such as gliders in the Game of Life are isolated bodies of matter. Any set of laws that describes a continuous world such as our own will have a concept of energy, which is a conserved quantity, meaning it doesn't change in time. The energy of empty space will be a constant, independent of both time and position. One can subtract out this constant vacuum energy by measuring the energy of any volume of space relative to that of the same volume of empty space, so we may as well call the constant zero. One requirement any law of nature must satisfy is that it dictates that the energy of an isolated body surrounded by empty space is positive, which means that one has to do work to assemble the body. That's because if the energy of an isolated body were negative, it could be created in a state of motion so that its negative energy was exactly balanced by the positive energy due to its motion. If that were true, there would be no reason that bodies could not appear anywhere and everywhere. Empty space would therefore be unstable. But if it costs energy to create an isolated body, such instability cannot happen, because, as we've said, the energy of the universe must remain constant. That is what it takes to make the universe locally stable—to make it so that things don't just appear everywhere from nothing.

If the total energy of the universe must always remain zero, and it costs energy to create a body, how can a whole universe be created from nothing? That is why there must be a law like gravity. Because gravity is attractive, gravitational energy is negative: One has to do work to separate a gravitationally bound system, such as the earth and moon. This negative energy can balance the positive energy needed to create matter, but it's not quite that simple. The negative gravitational energy of the earth, for example, is less than a billionth of the positive energy of the matter particles the earth is made of. A body such as a star will have more negative gravitational energy, and the smaller it is (the closer the different parts of it are to each other), the greater this negative gravitational energy will be. But before it can become greater than the positive energy of the matter, the star will collapse to a black hole, and black holes have positive energy. That's why empty space is stable. Bodies such as stars or black holes cannot just appear out of nothing. But a whole universe can.

Because gravity shapes space and time, it allows space-time to be locally stable but globally unstable. On the scale of the entire universe, the positive energy of the matter *can* be balanced by the negative gravitational energy, and so there is no restriction on the creation of whole universes. Because there is a law like gravity, the universe can and will create itself from nothing in the manner described in Chapter 6. Spontaneous creation is the reason there is something rather than nothing, why the universe exists, why we exist. It is not necessary to invoke God to light the blue touch paper and set the universe going.

Why are the fundamental laws as we have described them? The ultimate theory must be consistent and must predict finite results for quantities that we can measure. We've seen that there must be a law like gravity, and we saw in Chapter 5 that for a theory of gravity to predict finite quantities, the theory must have what is called supersymmetry between the forces of nature and the matter on which they act. M-theory is the most general supersymmetric theory of gravity. For these reasons M-theory is the *only* candidate for a complete theory of the universe. If it is finite—and this has yet to be proved—it will be a model of a universe that creates itself. We must be part of this universe, because there is no other consistent model.

M-theory is the unified theory Einstein was hoping to find. The fact that we human beings—who are ourselves mere collections of fundamental particles of nature—have been able to come this close to an understanding of the laws governing us and our universe is a great triumph. But perhaps the true miracle is that abstract considerations of logic lead to a unique theory that predicts and describes a vast universe full of the amazing variety that we see. If the theory is confirmed by observation, it will be the successful conclusion of a search going back more than 3,000 years. We will have found the grand design.

GLOSSARY

Alternative histories • a formulation of quantum theory in which the probability of any observation is constructed from all the possible histories that could have led to that observation.

Anthropic principle • the idea that we can draw conclusions about the apparent laws of physics based on the fact that we exist.

Antimatter • each particle of matter has a corresponding antiparticle. If they meet, they annihilate each other, leaving pure energy.

Apparent laws • the laws of nature that we observe in our universe—the laws of the four forces, and the parameters such as mass and charge that characterize the elementary particles—in contrast to the more fundamental laws of M-theory that allow for different universes with different laws.

Asymptotic freedom • a property of the strong force that causes it to become weaker at short distances. Hence, although quarks are bound in nuclei by the strong force, they can move within nuclei almost as if they felt no force at all.

Atom • the basic unit of ordinary matter, consisting of a nucleus with protons and neutrons, surrounded by orbiting electrons.

Baryon • a type of elementary particle, such as the proton or neutron, that is made of three quarks.

Big bang • the dense, hot beginning of the universe. The big bang theory postulates that about 13.7 billion years ago the part of the universe we can see today was only a few millimetres across. Today the universe is vastly larger and cooler, but we can observe the remnants of that early period in the cosmic microwave background radiation that permeates all space.

Black hole • a region of space-time that, due to its immense gravitational force, is cut off from the rest of the universe.

Boson • an elementary particle that carries force.

Bottom-up approach • in cosmology, an idea that rests on the assumption that there's a single history of the universe, with a well-defined starting point, and that the state of the universe today is an evolution from that beginning.

Classical physics • any theory of physics in which the universe is assumed to have a single, well-defined history.

Cosmological constant • a parameter in Einstein's equations that gives space-time an inherent tendency to expand.

Electromagnetic force • the second strongest of the four forces of nature. It acts between particles with electric charges.

Electron • an elementary particle of matter that has a negative charge and is responsible for the chemical properties of elements.

Fermion • a matter-type elementary particle.

Galaxy • a large system of stars, interstellar matter, and dark matter that is held together by gravity.

Gravity • the weakest of the four forces of nature. It is the means by which objects that have mass attract each other.

Heisenberg uncertainty principle • a law of quantum theory stating that certain pairs of physical properties cannot be known simultaneously to arbitrary precision.

Meson • a type of elementary particle that is made of a quark and an anti-quark.

M-theory • a fundamental theory of physics that is a candidate for the theory of everything.

Multiverse • a set of universes.

Neutrino • an extremely light elementary particle that is affected only by the weak nuclear force and gravity.

Neutron • a type of electrically neutral baryon that with the proton forms the nucleus of an atom.

No-boundary condition • the requirement that the histories of the universe are closed surfaces without a boundary.

Phase • a position in the cycle of a wave.

Photon • a boson that carries the electromagnetic force. A quantum particle of light.

Probability amplitude • in a quantum theory, a complex number whose absolute value squared gives a probability.

Proton • a type of positively charged baryon that with the neutron forms the nucleus of an atom.

Quantum theory • a theory in which objects do not have single definite histories.

Quark • an elementary particle with a fractional electric charge that feels the strong force. Protons and neutrons are each composed of three quarks.

Renormalization • a mathematical technique designed to make sense of infinities that arise in quantum theories.

Singularity • a point in space-time at which a physical quantity becomes infinite.

Space-time • a mathematical space whose points must be specified by both space and time coordinates.

String theory • a theory of physics in which particles are described as patterns of vibration that have length but no height or width—like infinitely thin pieces of string.

Strong nuclear force • the strongest of the four forces of nature. This force holds the protons and neutrons inside the nucleus of an atom. It also holds together the protons and neutrons themselves, which is necessary because they are made of still tinier particles, quarks.

Supergravity • a theory of gravity that has a kind of symmetry called supersymmetry.

Supersymmetry • a subtle kind of symmetry that cannot be associated with a transformation of ordinary space. One of the important implications of supersymmetry is that force particles and matter particles, and hence force and matter, are really just two facets of the same thing.

Top-down approach • the approach to cosmology in which one traces the histories of the universe from the "top down", that is, backwards from the present time.

Weak nuclear force • one of the four forces of nature. The weak force is responsible for radioactivity and plays a vital role in the formation of the elements in stars and the early universe.

ACKNOWLEDGEMENTS

THE UNIVERSE HAS A DESIGN, and so does a book. But unlike the universe, a book does not appear spontaneously from nothing. A book requires a creator, and that role does not fall solely on the shoulders of its authors. So first and foremost we'd like to acknowledge and thank our editors, Beth Rashbaum and Ann Harris, for their near-infinite patience. They were our students when we required students, our teachers when we required teachers, and our prodders when we required prodding. They stuck with the manuscript, and did it in good cheer, whether the discussion centred around the placement of a comma or the impossibility of embedding a negative curvature surface axisymmetrically in flat space. We'd also like to thank Mark Hillery, who kindly read much of the manuscript and provided valuable input; Carole Lowenstein, who did so much to help with the inside design; David Stevenson, who guided the cover to completion; and Loren Noveck, whose attention to detail has saved us from some typos we would not like to have seen committed to print. To Peter Bollinger: much gratitude for bringing art to science in your illustrations, and for your diligence in ensuring the accuracy of every detail. And to Sidney Harris: thank you for your wonderful cartoons, and your great sensitivity to the issues facing scientists. In another universe, you could have been a physicist. We are also grateful to our agents, Al Zuckerman and Susan Ginsburg, for their support and encouragement. If there are two messages they consistently provided, they were "It's time to finish the book already," and "Don't worry about when you'll finish, you'll get there eventually." They were wise enough to know when to say which.

And finally, our thanks to Stephen's personal assistant, Judith Croasdell; his computer aide, Sam Blackburn; and Joan Godwin. They provided not just moral support, but practical and technical support without which we could not have written this book. Moreover, they always knew where to find the best pubs.

INDEX

Page numbers of illustrations appear in italics.

ABOUT THE AUTHORS

STEPHEN HAWKING held the post of Lucasian Professor of Mathematics at Cambridge, the chair held by Isaac Newton in 1663, for thirty years. Professor Hawking is now Director of Research for the Centre for Theoretical Cosmology at the University of Cambridge. He has over a dozen honorary degrees, and was awarded the Companion of Honour in 1989. He is a fellow of the Royal Society and a member of the US National Academy of Science. His books include the bestselling *A Brief History of Time*, *Black Holes and Baby Universes and Other Essays*, *The Universe in a Nutshell* and *A Briefer History of Time*. Stephen Hawking is regarded as one of the most brilliant theoretical physicists since Einstein. He lives in Cambridge.

LEONARD MLODINOW is a physicist at Caltech University and the bestselling author of *The Drunkard's Walk: How Randomness Rules Our Lives*, *Euclid's Window: The Story of Geometry from Parallel Lines to Hyperspace*, and *Feynman's Rainbow: A Search for Beauty in Physics and in Life*. He also wrote for *Star Trek: The Next Generation*. He lives in South Pasadena, California.